OPTIMAL CONTROL OF SPATIAL SYSTEMS

The London Research Series in Geography

Geography is a wide-ranging field of knowledge, embracing all aspects of the relations between people and environments. This series makes available work of the highest quality by scholars who are, or have been, associated with the various university and polytechnic departments of geography in the London area.

One of geography's most salient characteristics is its close relationships with virtually all the sciences, arts and humanities. Drawing strength from other fields of knowledge, it also adds to their insights. This series highlights these linkages. Besides being a vehicle for advances within geography itself, the series is designed to excite the attention of the wider community of scholars and students. To this end, each volume is chosen, assessed and edited by a committee drawn from all the London colleges and the whole range of the discipline, human and physical.

1 *The land of France 1815–1914*
Hugh Clout

2 *Living under apartheid*
David Smith (editor)

3 *London's Green Belt*
Richard Munton

4 *Urban hospital location*
Leslie Mayhew

5 *Nature's ideological landscape*
Kenneth Olwig

6 *Optimal control of spatial systems*
K. C. Tan and R. J. Bennett

7 *Geomorphological hazards in Los Angeles*
R. U. Cooke

8 *Land use and prehistory in south-east Spain*
John Thornes and Antonio Gilman

OPTIMAL CONTROL OF SPATIAL SYSTEMS

K.C. Tan

Department of Economics and Statistics, National University, Singapore

R.J. Bennett

Department of Geography, University of Cambridge

London
GEORGE ALLEN & UNWIN
Boston Sydney

**George Allen & Unwin (Publishers) Ltd,
40 Museum Street, London WC1A 1LU, UK**

George Allen & Unwin (Publishers) Ltd,
Park Lane, Hemel Hempstead, Herts HP2 4TE, UK

Allen & Unwin Inc.,
9 Winchester Terrace, Winchester, Mass. 01890, USA

George Allen & Unwin Australia Pty Ltd,
8 Napier Street, North Sydney, NSW 2060, Australia

First published in 1984

British Library Cataloguing in Publication Data

Tan, K. C.
Optimal control of spatial systems.
1. Control theory 2. Mathematical optimization
I. Title II. Bennett, R. J.
629.8′312 QA402.3
ISBN 0-04-519018-6

Library of Congress Cataloging in Publication Data

Tan, K. C.
Optimal control of spatial systems.
Bibliography: p.
Includes index.
1. Spatial systems. 2. Control theory. I. Bennett, R. J. (Robert John) II. Title.
QA402.T36 1984 515 83-25701
ISBN 0-04-519018-6

Set in 10 on 12 point Times by Preface Ltd, Salisbury, Wilts.
and printed in Great Britain by Butler & Tanner Ltd, Frome and London

Preface and acknowledgements

This book is concerned with the theory and application of the methods of optimal control to dynamic systems which are distributed across an array of spatial zones or regions. Many statistical and mathematical developments have already been made with such spatial systems, but this book represents the first attempt to draw together the methods for *control* of spatial systems. The presentation begins with a simple introduction to spatial problems and the motivation for spatial control. This is then followed by Part I of the book which deals with each of the major aspects of the *theory* of spatial control systems. The simplest cases are of deterministic and stochastic systems with known parameter structure. These are discussed in Chapter 2. Chapter 3 introduces the situation in which there are linear equality and inequality constraints on the control variables. Chapters 4 and 5 introduce situations in which system parameters are unknown and are, respectively, either constant or time-varying. The theoretical content is concluded in Chapter 6 with a discussion of spatial control for cases in which both state and control variables are subject to constraints.

One of the main motivations behind the theoretical developments of constrained control problems undertaken in this book lies in the prospect of formulating as a control problem the distribution of central government grants between local authorities. This is one of the main applications discussed in Part II of the book. Other applications which are discussed at length are the control of national settlement systems and the control of manpower levels in various skill groups. Throughout the discussion there are simple worked examples which help the reader to develop understanding of the methods.

The development of a theory for control in constrained systems is one of the major contributions of this book. Most previous work has dealt only with unconstrained cases. When attempts are made to solve the optimal control problem subject to constraints, most existing methods yield at best only local optima. The improvement over existing methods given in this book yields truly optimal solutions in the deterministic case. This improved solution technique may also be used in situations where the control is to be carried out on system models where uncertainties are present. Although it does not give the true optimal solution in this case, it is optimal subject to various simplifying assumptions. Two kinds of uncertainties are considered. First, when the system model contains an additive random term (stochastic control problems) and second, when the model parameters are unknown. The first case is an easier problem to solve than the second, but solutions to both are given below. The book is aimed at a wide audience in economics, geography, planning, engineering, environmental science, operations research and regional science. The level and approach will make this an advanced reader in control theory in undergraduate courses and an important reference text elsewhere. This will be found all the more readily possible since the notation and approach follows closely that of two complementary books: R. J. Bennett (1979) *Spatial time series* (London: Pion) and R. J. Bennett and R. J. Chorley (1978) *Environmental systems* (London and New York: Methuen).

In writing this book we would wish to acknowledge the very considerable debt that we owe to a multitude of individuals and organisations. Much of the earlier work for this book was undertaken when both authors were at the Department of Geography, University College London. We also would like to express our grateful thanks to the Association of Commonwealth Universities who supported K. C. Tan during this work and to former colleagues in London, especially to Peter Wood who read many of the early drafts of much of the material, and to W. R. Mead. The later work for the book was undertaken at the University of Cambridge, the University of Science, Malaysia, and at the National University of Singapore. Part of the empirical work on the allocation of intergovernmental grants was made possible by the support of the British Social Science Research Council. To each of these institutions the authors express their sincere thanks. Valuable comments on the work were also received from Alan Wilson, for which we offer very great thanks. The manuscript was typed in draft by Mrs C. John and in final form by Mrs J. Ashman; cartography was undertaken by Arthur Shelley and Alick Newman. We thank the Institute of British Geographers for permission to reproduce Figure 7.4.

K. C. TAN
Singapore
August 1983

R. J. BENNETT
Fitzwilliam College
Cambridge

Contents

1 *Optimal control of spatial systems*

1.1 The purpose of control in spatial systems

Optimal control theory concerns the choice of disposition of a set of control instruments to achieve a given objective function, taking account of the dynamic properties of the system under consideration. The application of optimal control theory to spatial systems concerns the choice of disposition of a set of instruments or policy variables affecting different spatial locations to achieve a given objective function which is spatially variable so as to satisfy a set of system properties which are dynamic and which are evolving and adjusting over space.

Common control problems are the optimal regulation of water level in a reservoir, timing of traffic lights at a road junction, stabilisation of cyclical swings in economies, achievement of acceptable levels of clean air or pollution in rivers and lakes, etc. Common spatial control problems are control of pollution flux, the allocation of service outlets to markets, the disposition of spatial economic stabilisation devices, federal revenue and expenditure regulation, and regional policy.

The characteristics of the control problem are shown in Figure 1.1. The system which it is desired to control is represented by a model which is affected by two sets of system inputs: an *uncontrolled* input sequence (x_t) and an additional *controlled* input sequence (x_t^c). The controlled input sequence represents a variable that can be manipulated to influence the system under study and to induce the system output to approach more closely to 'desired' output settings. Variables amenable to such manipulation are termed policy *instruments*. Hence control systems are very like 'free' systems except that the input can be specified by the analyst, by a decision maker or by a political process. Control systems also introduce the concept of system goals. Any given output can be compared with a *reference, goal* or *target* (a_t) which represents the desired output level of the system. The difference between the real and desired output generates an 'error' sequence, which is then weighted by a chosen *performance index* or objective function which scales deviations from the desired system response. This is then used by a system controller to manipulate those variables which are amenable to control. This approach covers a wide variety of control systems that have been treated in economics, especially by Tinbergen (1937), in control engineering by Kalman (1958, 1960) and in environmental systems by Bennett and Chorley (1978).

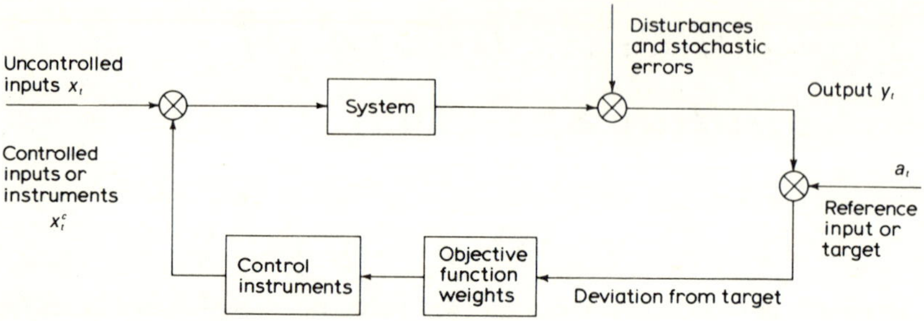

Figure 1.1 The configuration of a typical control system.

The control system shown in Figure 1.1 is a form of feedback device and can readily be automated once each of the elements a_t, x_t and the performance indicator is specified. Hence it is a control device much used in engineering and other physical systems in which the system inputs, outputs and behaviour are relatively predictable and are only slowly changing. In socio-economic and environmental systems, however, control must be implemented in conditions in which inputs, outputs and system behaviour are known with only low certainty, and in which changes in system behaviour resulting from learning and adaptation are the rule rather than the exception. In such systems we require more open and receptive control devices that are not rigidly constrained by feedback schemes. An alternative form of control device for such systems is shown in Figure 1.2. This differs from the automated feed-

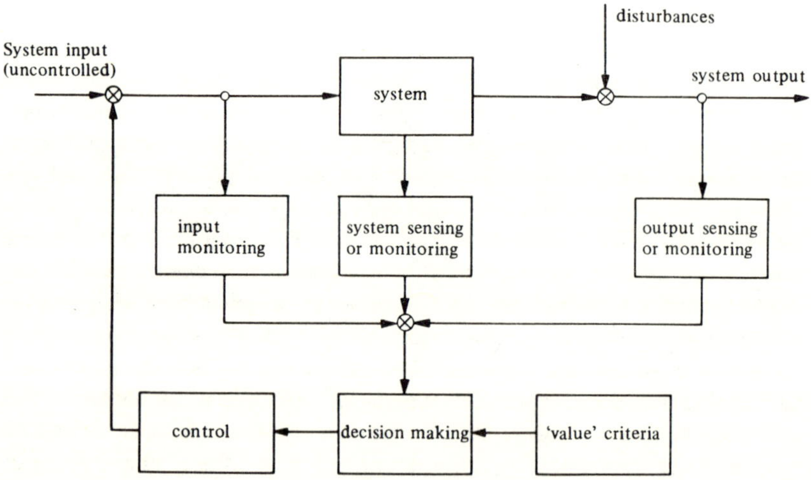

Figure 1.2 An expansion of the structure of control systems to show the way in which monitoring and decision making enter the control process.

back scheme in that changes in system input, output and transfer function are continuously monitored. This combined information is then used within a decision-making framework, which may be automated but will more usually be based upon a subjective set of judgments aided by quantitative assessments.

Most previous discussions of optimal control in spatial systems have been based upon models described by partial differential and difference equations (see, for example, Butkovskii 1969, Ahmed & Teo 1981). For such models it must be assumed that the space is continuously differentiable or at least regularly partitioned (e.g. on a Cartesian lattice) and that interaction between locations is controlled by a pattern of distance decay. In addition it is usually assumed that at least the structure (but not necessarily the parameters) of the model is known *a priori*. Such spatial models described by partial differential equations are of little utility for developing control models of regional economies, and this book is primarily concerned with extending the results of optimal control theory to situations in which the spatial domain of interest (e.g. a national economy) is partitioned into regions which interact arbitrarily such that no simple distance metric is necessarily involved.

The control of systems which are partitioned into a set of n spatial regions presents some new and fascinating problems in optimal control theory.

1.2 Categories of control problem

The control problem can be stated in general terms as the maximisation of an objective function of system behaviour that uses a given set of control instruments over a specified period of time. However, within this broad class of control problems, a wide variety of categories exists. The category of control solution developed depends on the nature of system disturbances (varying from minor random shifts to major structural shifts in the nature of the system under control), the nature of the objective function (linear, non-linear, quadratic, etc.), the time period over which it is sought to apply control (one-stage, T-stage, infinite), and the nature of the control strategy adopted (optimal or sub-optimal).

Corresponding to each category of control is a different control problem. Figure 1.3 displays a range of control methods in common use:

(a) *Uncontrolled systems*. These correspond to systems in which the inputs into the system are either not known or are not available for manipulation as policy instruments.

(b) *Open-loop control systems*. On the basis of available experience a control rule is present, as with traffic lights at a road intersection.

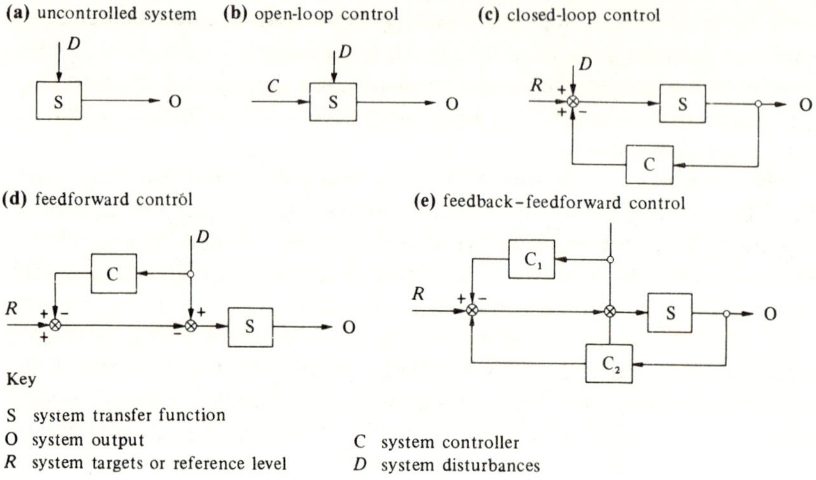

Figure 1.3 Categories of control system.

This system can be adjusted only infrequently and tends to ignore unexpected fluctuations.

(c) *Closed-loop control systems.* This is the most common category of control strategy and various categories of closed-loop control are available. Each requires the specification of a reference input or target, a comparison of the real system output and the target, and a *feedback* to counteract the effect of deviations. This deviation is then weighted to achieve optimum cancellation within a given time period. Two subcategories of closed-loop control are in common use: (i) *the regulator*, in which the targets are set at specified values for all time, and (ii) *the servomechanism*, in which the targets are varying functions of time.

(d) *Feedforward control systems.* Closed-loop systems are feedback devices which have the disadvantage of responding to system disturbances only after they have occurred. Feedforward control systems offer the advantages of detecting disturbances before they affect system operations. Control action is then initiated to counteract disturbances as they affect the system. This allows the theoretical possibility of perfect control if the disturbances can be anticipated perfectly.

(e) *Feedback–feedforward control systems.* Feedforward control suffers from the disadvantage that there is no comparison of the system output with a target to assess how far the control has been effective. Feedforward schemes are therefore seldom used in isolation. Combination with a feedback controller, however, allows the combination of the advantage of both schemes: anticipation of system disturbances and feedback to counteract imperfections in the feedforward scheme. This is the type of control system in most frequent use in industrial applications, and is also the most useful in socioeconomic and environmental control if it is combined with adaptive elements and with decision-making strategies (see Bennett & Chorley 1978).

In addition to these broad categories, the control problem can be solved by both *sub-optimal* and *optimal* control. In sub-optimal control the decision rule is chosen on some informed but arbitrary basis, which will not normally yield complete cancellation of input disturbances. Optimal control, however, can achieve complete elimination of any disturbances from targets.

Sub-optimal control systems Sub-optimal control is an *ad hoc* method in which various rules are applied and control settings are made on the basis of observed disturbance properties. For feedback schemes, such control systems are given simply by

$$x_t = -k(y_t - a_t) \tag{1.1}$$

where x_t, the system input, is a control instrument which is set using the deviation of the system output y_t from the target a_t weighted by a set of *control parameters k*. Various strategies can be built into such control strategies, but no consideration is given to control costs or to the optimality of cancellation of the disturbances. The main strategies possible each involve a different choice of k in Equation (1.1).

(a) *On–off action* (bang-bang control):

$$x_t = \begin{cases} -k_1(y_t - a_t), & |y_t - a_t| \leqslant \alpha \\ -k_2(y_t - a_t), & |y_t - a_t| > \alpha \end{cases} \tag{1.2}$$

(b) *Proportional action* (P):

$$x_t = k_{\mathrm{P}}(y_t - a_t) \tag{1.3}$$

(c) *Integral action* (I):

$$x_t = -k_{\mathrm{I}} \sum_{t=0}^{T} (y_t - a_t) \tag{1.4}$$

(d) *Derivative action* (D):

$$x_t = -k_{\mathrm{D}}[(y_t - a_t) - (y_{t-1} - a_{t-1})] \tag{1.5}$$

On–off action is the simplest form of control which switches between two control settings depending upon the magnitude of the system deviation from target. Often the first constant k_1 is set to zero, so that no control action is taken if deviations are below a certain value. This can be extended to multipositional control settings k_i, ($i = 1, 2, \ldots$). Proportional action is a simple generalisation of the multipositional and on–off cases, where the magnitude of the feedback gain is a simple linear relation to the magnitude of the deviation. Integral action gives a feedback control setting proportional to the sum of deviations over the previous time periods; and derivative action gives a gain proportional to the rate of change of the deviations over the previous time period. The most frequent form of *ad hoc* control is a combination of several of these control actions. Common examples are proportional-plus-integral (PI) or two-term controller and proportional-plus-integral-plus-derivative (PID) control or three-term controller.

Optimal control systems Although sub-optimal controllers of the PID type can provide extremely effective strategies, this book is concerned essentially with optimal control strategies. The difficulty in applying sub-optimal control systems is that there are usually few *a priori* grounds for choosing one particular control strategy rather than another. Much practical economic and environmental control has followed such strategies and has, as a result, been rather hit and miss. It is often preferable, therefore, to use the available knowledge of the system properties to design the control strategy.

Optimal control strategies are derived by recognising that if the structure of the system of interest is known, then an optimal controller can be determined simply by inverting its system equation. Thus, given a system equation $y_t = Sx_t$, the optimal controller can be derived by simply inverting the system equation to give

$$x_t = -S^{-1}(y_t - a_t) = -k(y_t - a_t) \tag{1.6}$$

When the system is deterministic, this is referred to as a *dead-beat* controller, whilst for stochastic systems it is referred to as a *minimum-variance* controller.

The simplest forms of optimal control device simply use the inversion property of Equation (1.6); then the system coefficients and constants are used as weights of deviations of outputs from targets. In many applications, however, the properties of the system are too complex, too uncertain or too unstable to allow this simple inversion to be achieved. In other applications, although it may be technically feasible simply to invert the system equation, the resulting control device is too costly to be implemented. Such practical constraints stimulate the major class of control problems which are the main concern of this book. These are optimal control problems, which use the properties of the system equation, but which also impose constraints on the system outputs, the controls or both. The constraints in all cases are forms of 'costs' which act as weights. The combination of weights on outputs, or controls or both outputs and controls defines a *performance criterion* for optimal control (often also referred to as the *objective function* or *figure of merit*).

A major problem in defining an optimal control strategy is in the choice of the most appropriate performance criterion. Denoting this by $W(\cdot)$, the major forms of criteria usually employed are summarised in Table 1.1. These criteria differ in terms of the time period over which it is sought to cancel destinations, and in the weighting given to errors from targets. The control action deriving from these criteria differs most markedly between one-stage and T-stage methods. In one-stage control it is attempted to bring deviations from targets to a minimum by the next sampling instant, but for T-stage control it is assumed that a period of length T is available in which to minimise deviations from target, that is, an extended planning time horizon. The most usual weightings given to errors from targets are those of minimum error and least squares. The criteria W_2 and W_5, based on least squares, have the useful property that the optimal control is the *dual* of the least-squares parameter estimation problem (Kalman 1958, 1960) and is a linear function of the system

Table 1.1 The main categories of control performance criteria and their objective functions.

(1) One-stage minimum error

$$W_1 = E(y_t - a_t)$$

(2) One-stage minimum squared error

$$W_2 = E[(y_t - a_t)^2]$$

(3) One-stage absolute error

$$W_3 = E|y_t - a_t|$$

(4) T-stage integral error

$$W_4 = \lim_{T\to\infty} \frac{1}{T}\int_0^T [y(t) - a(t)]\,dt \approx E\left(\frac{1}{T}\sum_{t=1}^{T} y_t - a_t\right)$$

(5) T-stage minimum squared error

$$W_5 = \lim_{T\to\infty} \frac{1}{T}\int_0^T [y(t) - a(t)]^2\,dt \approx E\left(\frac{1}{T}\sum_{t=1}^{T} (y_t - a_t)^2\right)$$

(6) T-stage minimum absolute error

$$W_6 = \lim_{T\to\infty} \frac{1}{T}\int_0^T |y(t) - u(t)|\,dt \approx E\left(\frac{1}{T}\sum_{t=1}^{T} |y_t - a_t|\right)$$

output. The solutions using the Laplace absolute deviation criteria W_3 and W_6, however, yield non-linear matrix Riccati equations (Bryson & Ho 1969) and it is necessary to integrate both the model and control over the T steps of the planning time period.

In practice, most optimal control solutions employ a combined performance index or objective function which incorporates two elements: (a) the cost of deviations from targets and (b) costs of employing a given control instrument. A common choice is the *quadratic* control criterion:

$$W_2 = E(M_1(y_t - a_t)^2 + M_2 x_t^2) \tag{1.7}$$

where M_1 and M_2 are weighting constants (or matrices in the multivariate case) which determine respectively the relative costs of deviations from targets and the costs of control. It is often especially difficult in socio-economic and environmental systems to decide on appropriate values for the objective function terms M_1 and M_2. In particular, Bennett and Chorley (1978, Ch. 4) note the need for a capacity to deal with singularities (saddles and branching points) in the control space which may lead to catastrophic 'flips' from one regime to another.

Despite its simplicity, the quadratic control criterion has particularly useful mathematical properties that make it invaluable in practice. The most important of these is that the presence of stochastic disturbances does not complicate optimal control solutions unduly since the optimal quadratic controller will often have the same form in the stochastic as in the deterministic case. This is due to the *certainty equivalence principle* (Simon 1956, Theil 1957), stated by Simon as follows: when a quadratic cost criterion (W_2 or W_5) is used in the control problem under uncertainty, then only the expectation and not the probability density function (pdf) of the control variables need be known. Hence, the minimum variance mean square error forecasts can be used in the control solution to provide information to feedforward control devices, provided that they are unbiased (Holt *et al*. 1960); that is, the mean value of all systems variables can be used. For this reason the quadratic control criterion is used exclusively in this book.

1.3 *The state of the art in spatial systems control*

The art of optimal control theory is a comparatively new one, and it was only during the late 1930s and 1940s that the necessary theoretical advances were made which allowed many control problems to be solved. First in importance of these developments ranks the reinterpretation of optimal filtering and estimation theory, developed by Kolmogorov (1941) and Wiener (1949), into the discrete domain by Kalman (1960).

Secondly, was the recognition of the control problem as the inverse of the estimation problem.

For the control problem two apparently different approaches were available – Pontryagin's minimum principle (Pontryagin *et al.* 1962) and dynamic programming. On the one hand, the Pontryagin *minimum principle* was based on the calculus of variations, was appropriate only for deterministic control problems and was originally developed to solve continuous rather than discrete time problems. Several attempts had been made to derive the discrete-time minimum principle, e.g. Rosen (1967), Dorfman (1969) and Fan and Wang (1964). Chow (1975), for example, points out the similarity of the Pontryagin results with the method of Lagrange multipliers, and Rosen shows that the discrete optimal control problem may be treated as a larger static convex programming problem via the Kuhn–Tucker theory of non-linear programming (see also Pindyck 1973). This result is used for deriving solutions to constrained optimal control problems in forthcoming chapters. On the other hand, *dynamic programming* allowed solution of the multi-stage control problem by working backwards one period at a time from the final to the initial state. This method relies on Bellman's principle of optimality. This states that the optimal control policy has the property that, whatever the current conditions are, the optimal control decisions for the remaining stages are independent of the control decisions taken in previous stages (Bellman 1957). It is this principle which enabled T-stage control problems to be solved as a sequence of a single-stage problem.

The linkage between dynamic programming and the Pontryagin minimum principle provides the basis for the development of modern control theory. A major motivation behind this development was the space exploration programme during the 1950s and 1960s. The optimal control problem in this case was to enable a predetermined minimum fuel flightpath of a rocket to be calculated and to achieve successful navigation towards reaching the prespecified target (e.g. the Moon). Despite this specific stimulus, the very general nature of optimal control problems has not prevented these new techniques from finding relevance and usefulness in other areas, e.g. management of environmental pollution, migration, urban and regional planning, intergovernmental resource allocation, water management in catchments, and location–allocation models of public facilities, to mention but a few.

An early major area of application of control theory techniques in spatial systems is provided in the planning of population levels in national settlement systems. For example, MacKinnon (1975) discusses interregional migration models of a Markovian type within a control theory framework. When the objective function is non-linear, several attempts have been suggested to yield feasible solutions. For the quadratic criterion function, Mehra (1975) provides an heuristic method based on the minimum principle with an added branch and bound algorithm. Another method suggested by Evtushenko and MacKinnon (1976) is the penalty function technique of non-linear programming, which allowed inclusion within the objective function of positive terms containing the constraints of the problem (Zangwill 1969), and this has been extended by Propoi and Willekens (1978) and Nijkamp (1980).

In the wider context of urban and regional planning, the application of optimal control methods has been widely discussed (e.g. Domanski 1973, Isard & Liossatos 1975, Bennett & Chorley 1978, Bennett & Tan 1978, Lee 1978, Williams & Wilson 1978, Bennett 1979, Nijkamp 1980, Wilson 1981, Wilson & Bennett 1984). Using Pontryagin's minimum principle, Domanski considers the problem of optimal economic growth within a system of regions. Isard and Liossatos' discussion involves the optimal distribution of consumption for space–time models under different specifications of a production function. On a general level, the use of control theory methods as a tool for the dynamic analysis of urban and regional planning models is highlighted by Lee (1978), Williams and Wilson (1978) and Bennett and Chorley (1978). Bennett and Tan (1978) extend these applications to regional planning problems with an uncertain model. In addition, Williams and Wilson (1978), Coelho and Wilson (1976), Phiri (1979) and Nijkamp (1980) have applied optimal control theory to the allocation of retail outlets and housing stock in urban systems models. Some other examples of applications are given by Felderer (1975), Tabata (1976) and Karlqvist *et al.* (1978). Felderer considers the problem of resource allocation in regional educational planning. The model arrived at is based on the desire to optimise over time and over spatial regions, and the objective function is formulated as the quadratic. In finding the solution, the problem is actually static and the non-linear objective function is then linearised, thus giving a large linear programming problem. The application considered by Tabata concerns the problem of optimal investment allocation in a system of n regions and is an extension of earlier work with just two regions and centrally controlled investments (cf. Raman 1963). Karlqvist *et al.* develop, with special reference to southeastern Australia, a mathematical regional planning model.

Nijkamp (1976, 1980), Nijkamp and Verhage (1976) and Sakawa and Ueda (1978) have suggested application of optimal control theory to environmental pollution problems. Although these studies are concerned with the control of pollution levels by investment policies, they differ in their analysis. Nijkamp accounts for the regional aspects of the problem, while the approach of Sakawa and Ueda relates environment pollution to economic growth. The minimum principle is used to derive the optimal control solutions in both instances. In addition, Nijkamp

(1979a,b, 1981) and Nijkamp and Spronk (1980) have extended these methods to cover 'soft' data and multicriteria problems.

The application of optimal control theory to non-spatial systems in economics dates back much earlier and the range of problems considered is indeed vast. There are several survey articles on these different applications, e.g. Athans and Kendrick (1974), Fair (1974), Kendrick (1976, 1977), Ando *et al.* (1978), and Committee on Policy Optimisation (1978). Almost all of the major economic applications use control theory to design an economic stabilisation policy. The first attempts to apply control theory to economics were those by Tustin (1953), Phillips (1954) and Theil (1964), who discuss the regulation of deterministic macro-economic models based on transfer function representations. This early work did not use the state-space form. Early discussions of optimal economic stabilisation policy with models represented in state-space form are those by Pindyck (1973) and Turnovsky (1973). Pindyck's is a major empirical study demonstrating the use of optimal control theory with a small quarterly linear model of the US economy. Other important applications are those by Sengupta (1970), Howarth and Parks (1971), Brito (1973), Holbrook (1974), Vishwakarma (1974), Chow (1975, 1977, 1981), Aoki (1976), Pitchford and Turnovsky (1977), Turnovsky (1977a,b), Kamien and Schwartz (1981) and Tzafestas (1982). A further set of applications concerns the consistency of optimal economic policies in aggregate monetary policy, e.g. Marty (1968), Sargent and Wallace (1973), Brock (1974, 1975), Calvo (1978a,b) and Turnovsky (1978); while Diamond and Mirlees (1971a,b) and Turnovsky and Brock (1980) have assessed optimal tax policies.

In the analysis of environmental systems, major discussions of hydrological control are provided by Bennett and Chorley (1978, pp. 319–31, 342–52 and 356–68), control of river and air pollution is discussed by Nijkamp (1975), Young (1978), Whitehead and Young (1978) and Whitehead *et al.* 1979), and control of plant growth and irrigation is discussed by Lai (1979).

Although many early applications required deterministic models, an important result was obtained by Simon (1956) and Theil (1957) who demonstrated that the optimal control settings of linear models with an additive random term can be found in the same way as deterministic models. As noted earlier, this result is known as the *principle of certainty equivalence* and requires a quadratic objective function. Interest in applying optimal stochastic control was given another boost with the development of the *Kalman filter*, an optimal recursive estimation scheme (Kalman 1960, Kalman & Bucy 1961), made possible by the representation of dynamic models in state space form. Vishwakarma (1974) and Chow (1975, 1981) show how the Kalman filter may be used when a stochastic linear dynamic model is to be controlled but the variables can only be observed indirectly. Peterka and Åstrom (1973) extend this to multivariate systems. In these situations the determination of the optimal control rule may be found according to the *separation theorem*, which states that the estimation of the unknown state variables and the setting of the optimal control may be done as two independent steps. However, applications to macro-economic planning have nearly always assumed that the system variables are directly measurable (see Chow 1973, Sorenson 1976).

The application of optimal control theory is not restricted to linear models alone. For example, a number of studies have considered the control of *non-linear* macro-economic dynamic models. Using the method of dynamic programming, the control of non-linear models with known and unknown parameters have been discussed by Chow (1975, 1976a, 1977), who recommends the approach of feedback control yielding the optimal control setting as a function of the previous system variables. The same feedback control approach has been proposed to handle large scale non-linear econometric models (Chow & Megdal 1978); and Holbrook (1974, 1975) has considered the same problem but with a deterministic non-linear model applied to the Canadian economy. These studies depend on first finding a suitable linear model to approximate the non-linear structure, and then proceeding in the analysis using the linear model. One technique of linearising the non-linear model by using a Taylor expansion was first suggested in economics by Theil (1964). Of course, such linear control solutions derived for non-linear systems are only approximately optimal.

A further set of control problems arise when either the control variables or the state variables of the system are subject to constraints such that they cannot be set arbitrarily to any value. Various strategies for solving such problems have been suggested. *Ad hoc* solutions are provided by appropriate choice of weights in the quadratic objective function. This is more an alternative to methods of imposing constraints directly (Chow 1975), and has been considered unsatisfactory by many (e.g. Chan & Maille 1975). Several 'engineering' attempts at solving the optimal control problem with constraints have been suggested, but these apply only to deterministic problems (Ritch 1973, Chan & Maille 1975, Abdelmalek 1978, Owens & Jones 1978). A more general approach to optimal control under constraints has been developed by Phiri (1979), Tan (1979a,b) and Lee (1982) and is discussed at length in Chapters 3–5 below.

A further problem in control theory, which has been less frequently addressed, derives from the case when system parameters are unknown and must be estimated. This results in an extremely difficult problem to solve since the separation theorem, discussed above, is invalid. Various approaches to this problem have been developed, but all rely on the result, due to Kalman (1960), that the control and estimation problems are *duals* of each other. As a result the Kalman filter is often applied to this problem (see, for example, Farison *et al*. 1967). Another method, known as the self-tuning regulator, due to Åstrom and Wittenmark (1973) and Peterka and Åstrom (1973), can be used in the case of constant but unknown system parameters. This assumes that the parameters are known, and then determines the control rule; the unknown parameters are then replaced in the expression for the control law by the appropriate estimates and the whole process repeated iteratively. The resulting control setting is asymptotically optimal in that the parameter estimates converge to the true parameter values. Such a control law is said to have the self-tuning property (Bertsekas 1976). However, these methods of estimating unknown parameters for optimal control have been developed by control engineers and have not been widely adopted for spatial policy analysis. Applications to macro-economics have instead been based on two methods due respectively to Chow (1975) and to MacRae (1975). The Chow method assumes that the constant but unknown parameters are initially estimated and these estimates are not revised when the system is being controlled. Alternatively the parameters can be 'learned' over time and this allows their estimates to be updated. The MacRae method incorporates learning using a Bayesian approach. Assumptions are made relating the posterior mean and covariance of the unknown parameters to the prior mean and covariance. To simplify the derivation of the control rule, MacRae replaces the unknown future observations by the predicted observations. In the restricted case, when parameter uncertainty is limited to the coefficients of the state variables, Kendrick and Majors (1974) discuss a similar method of augmenting the state vector to include the random parameters. Developments of these methods are discussed in later chapters.

1.4 Plan of the book

This book is concerned with developing a general body of theory and applications for control of spatially distributed systems. In each of the chapters a particular control situation is developed and a set of optimal control strategies are derived. Part I of the book deals with basic theory. In Chapter 2 the simplest case of an optimal control problem with a quadratic objective function and a linear dynamic model is discussed and solutions employing the method of dynamic programming are introduced. This chapter also includes discussion of the subject of stochastic control, exhibiting the certainty equivalence principle, the separation theorem, the Kalman filter and the representation of spatial systems suitable for control. Chapter 3 introduces problems in optimal control subject to constraints on the control variables. The method proposed yields the true optimal solution in the deterministic case. This chapter also demonstrates how the method may be used for systems with an additive disturbance term. Chapters 4 and 5 then consider the constrained control problem with unknown system parameters. The case of constant but unknown parameters is dealt with first in Chapter 4, then Chapter 5 considers the more general case of time-varying unknown parameters. In Chapter 6 the deterministic constrained control problem is again considered, but now with constraints on state variables as well as on the control variables.

Part II of the book is concerned with examining the application of the optimal control technique to a number of practical examples. The most complex of these is the use of optimal control methods for allocation of intergovernmental grants in England and Wales (the Rate Support Grant). This is discussed in Chapter 7. The following chapter then develops two smaller-scale applications to national settlement system planning and to national manpower planning. Each of the control problems analysed has characteristics in common, the most important being that there are constraints (both equality and non-negativity constraints) on the control variables. The treatment of spatial control problems in such situations has seldom before been tackled, and their solution represents an important theoretical contribution of this book.

PART I THEORY

2 *Spatial control systems*

2.1 *Spatial systems*

Spatial systems, as conceived in this book, consist of a set of n interacting, partially interacting or separate zones with dynamic properties which govern the behaviour of phenomena within each zone. Each zone may have one or many inputs and outputs, some of which may be controllable, whilst others may not.

An example of such spatial systems is a set of social and economic indicators which may be used to characterise different economic planning regions. These evolve over time with strong internal dynamic structures which derive from the nature of the industry and residential groups in each location. In addition to internal dynamics, however, there are also interregional linkages which derive from the flow of products, capital, labour and raw materials between locations as a set of spatial interactions. As a result, models which seek to capture the dynamics of the spatial system require complex lead-lag specifications for their internal and external dynamics.

In the discussion below, this modelling problem is approached by specifying, first, the set of inputs into each of the n spatial zones in the area under study; secondly, the set of outputs from each of the n spatial zones; and thirdly, the set of input–output relations which not only link each region to each other in a form of spatial interaction, but also affect the internal dynamics in each region. This results in a so-called vector–matrix representation of the spatial system. The vector is the string of inputs or outputs for each region, and the matrix is the set of links with and between regions which couples inputs to outputs. It should be noted that this representation differs from the approach to spatial systems normally developed in engineering. There the treatment is usually in terms of partial differential equations (see, for example, Butkovskii 1969, Ahmed & Teo 1981). The vector–matrix form allows simpler mathematical and statistical solutions than these partial differential equations; and it is much better suited for most socio-economic and environmental systems where spatial continuity properties (e.g. of stationarity in space) do not usually hold. We start by defining this structure in mathematical form below.

For univariate spatial systems with n zones there will be n inputs. These may be described by the vector

$$\underset{(n\times 1)}{x_t} = \begin{bmatrix} x_{t1} \\ x_{t2} \\ \vdots \\ x_{tn} \end{bmatrix} \tag{2.1}$$

where the subscript ($j = 1, 2, \ldots, n$) is used to define the input into the jth region. From this any time series of spatial inputs will be described by the term $\{x_t\}$ as the vector sequence over the n locations. In this and subsequent definitions we follow closely the notation employed in Bennett (1979).

The output of a univariate spatial system is similarly described by an n-dimensional vector:

$$\underset{(n\times 1)}{y_t} = \begin{bmatrix} y_{t1} \\ y_{t2} \\ \vdots \\ y_{tn} \end{bmatrix} \tag{2.2}$$

In both (2.1) and (2.2) it is assumed for the present that each of the n zones has an input and output, but cases of unobservable and uncontrollable systems may arise in practice.

The univariate system operation describing the relation between spatial inputs and outputs in the full case will be composed of $n \times n$ sub-elements which link each region to each other region, i.e.

$$\underset{(n\times 1)}{y_t} = \underset{(n\times n)}{S} \quad \underset{(n\times 1)}{x_t}$$

or

$$\begin{bmatrix} y_{t1} \\ y_{t2} \\ \vdots \\ y_{tn} \end{bmatrix} = \begin{bmatrix} S_{11} & S_{12} & \ldots & S_{1n} \\ S_{21} & S_{22} & \ldots & S_{2n} \\ \vdots & \vdots & & \vdots \\ S_{n1} & S_{n2} & \ldots & S_{nn} \end{bmatrix} \begin{bmatrix} x_{t1} \\ x_{t2} \\ \vdots \\ x_{tn} \end{bmatrix} \tag{2.3}$$

Some elements S_{ij} may be zero where no interaction or coupling between areas occurs. A typical structure of Equation (2.3) is shown in

Figure 2.1 for a two-region case. In practice each element S_{ij} will itself be composed of a number of subelements of system parameters.

For multivariate spatial systems the notation developed above can be extended to cases in which zones may have more than one input or output. For such cases with n zones the jth zone will have m_j inputs and n_j outputs. These can be defined by the vectors

$$\underset{((m1+m2+\ldots+mn)\times 1)}{x_t} = \begin{bmatrix} x_{t1}^1 \\ x_{t1}^2 \\ \vdots \\ x_{t1}^{m1} \\ \vdots \\ x_{tn}^1 \\ x_{tn}^2 \\ \vdots \\ x_{tn}^{mn} \end{bmatrix} \tag{2.4}$$

and

$$\underset{((n1+n2+\ldots+nn)\times 1)}{y_t} = \begin{bmatrix} y_{t1}^1 \\ y_{t1}^2 \\ \vdots \\ y_{t1}^{n1} \\ \vdots \\ y_{tn}^1 \\ y_{tn}^2 \\ \vdots \\ y_{tn}^{nn} \end{bmatrix} \tag{2.5}$$

There may, of course, be different numbers of inputs and outputs in each region, and hence the use of the subscripts n_j and m_j to denote these numbers. For the interaction of these n_j outputs and m_j inputs over the n regions there will be a system interaction matrix with elements S_{ij}^{kl}:

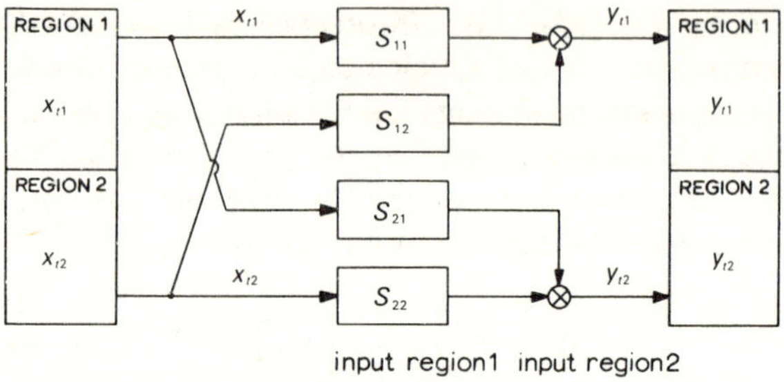

Figure 2.1 Representation of a two-region system in terms of the 2 × 2 transfer function operator S_{ij} (after Bennett & Chorley 1978, Fig. 4.2).

$$
\begin{bmatrix} y^{1}_{t1} \\ y^{2}_{t1} \\ \vdots \\ y^{n_1}_{t1} \\ \vdots \\ y^{1}_{tn} \\ y^{2}_{tn} \\ \vdots \\ y^{n_n}_{tn} \end{bmatrix}
=
\left[\begin{array}{cccc|cccc}
S^{11}_{11} & S^{12}_{11} & \cdots & S^{1m_1}_{11} & S^{11}_{1n} & S^{12}_{1n} & \cdots & S^{1m_n}_{1n} \\
\cdot & \cdot & & \cdot & \cdot & \cdot & & \cdot \\
\cdot & \cdot & & \cdot & \cdot & \cdot & & \cdot \\
\cdot & \cdot & & \cdot & \cdot & \cdot & & \cdot \\
S^{n_1 1}_{11} & S^{n_1 2}_{11} & \cdots & S^{n_1 m_1}_{11} & S^{n_1 1}_{1n} & S^{n_1 2}_{1n} & \cdots & S^{n_1 m_n}_{1n} \\
\hline
\vdots & & & & & & & \vdots \\
\hline
S^{11}_{n1} & S^{12}_{n1} & \cdots & S^{1m_1}_{n1} & S^{11}_{nn} & S^{12}_{nn} & \cdots & S^{1m_n}_{nn} \\
\cdot & \cdot & & \cdot & \cdot & \cdot & & \cdot \\
\cdot & \cdot & & \cdot & \cdot & \cdot & & \cdot \\
\cdot & \cdot & & \cdot & \cdot & \cdot & & \cdot \\
S^{n_n 1}_{n1} & S^{n_n 2}_{n1} & \cdots & S^{n_n m_1}_{n1} & S^{n_n 1}_{nn} & S^{n_n 2}_{nn} & \cdots & S^{n_n m_n}_{nn}
\end{array}\right]
\begin{bmatrix} x^{1}_{t1} \\ x^{2}_{t1} \\ \vdots \\ x^{m_1}_{t1} \\ \vdots \\ x^{1}_{tn} \\ x^{2}_{tn} \\ \vdots \\ x^{m_n}_{tn} \end{bmatrix}
\tag{2.6}
$$

The element S^{kl}_{ij} gives the interactions between output k in region i and the input l in region j. Again, each element S^{kl}_{ij} will be composed of a subset of system parameters governing dynamic properties.

This vector–matrix representation of spatial systems is quite distinct from the representation most commonly employed for distributed parameter systems. The more common representation is to invoke continuity relations between zones such that partial differential operators can be employed to compress the dimensionality of the system. This approach assumes the same structure of relations between any two zones, irrespective of their location, provided that they are the same distance apart. This assumes a form of spatial stationarity and isotropy. Such an assumption may be appropriate for many physical and engineering systems for which have been developed very successful control systems using the distributed parameter approach. However, the limitations imposed by this assumption in many environmental and socio-economic systems suggests that the more flexible representation defined in Equation (2.6) is more appropriate. This is the approach which, in various forms, is followed below. It does suffer from the drawback that the dimension of the resulting spatial models can be very large. However, there are usually methods available which can be employed to overcome this problem and the representation does offer considerable advantages in ease of manipulation, developments that can be made from this base, and the statistical estimation techniques that can be applied to it in cases where the system parameters are unknown.

The vector–matrix form also has the other advantages of allowing ready identification and estimation of Equations (2.3) or (2.6). This aspect is not the prime motivation for this book, but in the complementary volume (Bennett 1979) the simplicity of the vector–matrix form is argued to offer considerable advantages over comparable partial differential equations because of the relative ease of applying established techniques of identification and estimation. *Identification* refers to the step in the analysis which determines how many and what form of parameters make up the input–output relation S_{ij} in (2.3) or (2.6). In addition it concerns prior specification of the form and internal structure of the model. *Estimation* refers to a second step of analysis of determining the magnitude and sign (positive or negative) of the parameters. The two steps, therefore, serve to define the structure of the system state equation in its entirety. They are not without their own difficulties, and the reader is referred to the standard texts on how these two steps can be undertaken. In statistical estimation theory Brito (1973), Sorenson (1976), Bennett (1979), Paelinck and Klaassen (1979) and Wilson and Bennett (1984) cover the main topics. In urban planning applications the calibration and fitting methods which have been developed for complex sets of simultaneous and non-linear systems equations are described by Wilson (1970, 1974, 1981) and Batty (1976).

2.2 *Mathematical models of spatial control systems*

The discussion above has been developed in the context of the so-called *structural form*. Unfortunately such models are usually unsuitable for optimal control systems. One way to overcome this problem is to put structural form models into reduced form (see Pindyck 1973, Chow 1975, Pagan 1975, Turnovsky 1977a,b).

The difficulty with the structural form can be understood for spatial systems for the following two-region example, in which there is one variable in each region with interactions between regions given by the following equations:

$$y_{t,1} = a_1 y_{t-1,1} + a_2 y_{t2} + a_3 y_{t-1,2} + a_4 x_t$$
$$y_{t,2} = b_1 y_{t-1,2} + b_2 y_{t1} + b_3 y_{t-1,1} + b_4 x_t \tag{2.7}$$

where $a_1, \ldots, a_4, b_1, \ldots, b_4$ are the parameters of the model, $y_{t,i}$ denotes the endogenous variable at time t in region i, and x_t represents the exogenous variable (which affects both regions simultaneously) at time t. This model expresses the value of an endogenous variable in one region for the current time period t as a function of, first, the value taken by the lagged endogenous variable in the same region; secondly, the values taken in the other region by the current and lagged variable; and thirdly, the current exogenous variable. The interactions in the equation are simultaneous (y_{t2} determines y_{t1}, and y_{t1} determines y_{t2}). The example of the simultaneous equations (Eqs 2.7) is a structural form of model representation. Frequently lagged exogenous variables can also appear in this model. In this case the set of lagged endogenous variables and exogenous variables (current and lagged) are called predetermined variables. The model (2.7) may be represented diagrammatically as shown in Figure 2.2. This structural form creates a number of practical difficulties arising from the simultaneity of both y_t and y_{t-1} between regions. As a result, it is frequently necessary to express the current endogenous variables as functions of the predetermined variables. This eliminates the simultaneous effects given in (2.7) and the model is then said to be in *reduced form*. The reduced form is found by

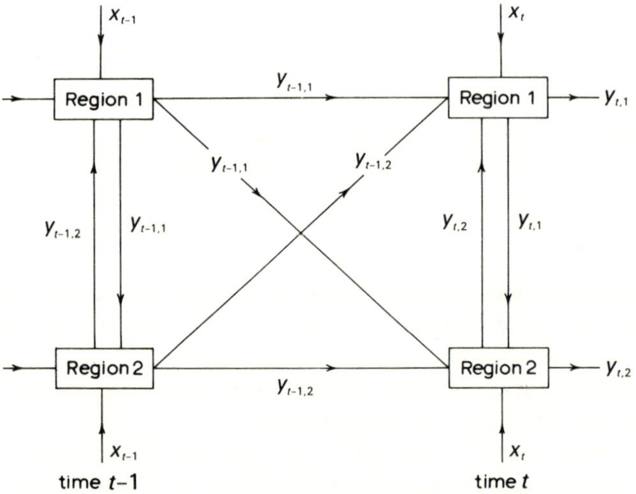

Figure 2.2 A two-region system exhibiting the structural form of equations.

solving the structural form for the current endogenous variable. For example, consider the reduced form representation of (2.7). This is given by

$$
\begin{aligned}
y_{t,1} &= \frac{1}{1-a_2b_2}[(a_1 + a_2b_3)y_{t-1,1} + (a_3 + a_2b_1)y_{t-1,2} \\
&\quad + (a_4 + a_2b_4)x_t] \\
y_{t,2} &= \frac{1}{1-a_2b_2}[(a_1b_2 + b_3)y_{t-1,1} + (a_3b_2 + b_1)y_{t-1,2} \\
&\quad + (a_4b_2 + b_4)x_t]
\end{aligned}
\tag{2.8}
$$

Notice that the coefficients of the reduced form (2.8) are all functions of the original structural parameters. The model (2.8) is shown diagrammatically in Figure 2.3.

The reduced form system of Equations (2.8) can be written more compactly in matrix notation. The matrix form permits a more convenient generalisation to an n-region spatial system. When there is also a general set of time lags for both endogenous and exogenous variables, such models may be defined in the general form

$$y_t = A_{1t}y_{t-1} + \ldots + A_{pt}y_{t-p} + B_{0t}x_t + \ldots + B_{qt}x_{t-q} \tag{2.9}$$

where $y_t = (y_{t1}, \ldots, y_{tn})'$, is an n-dimensional vector of endogenous

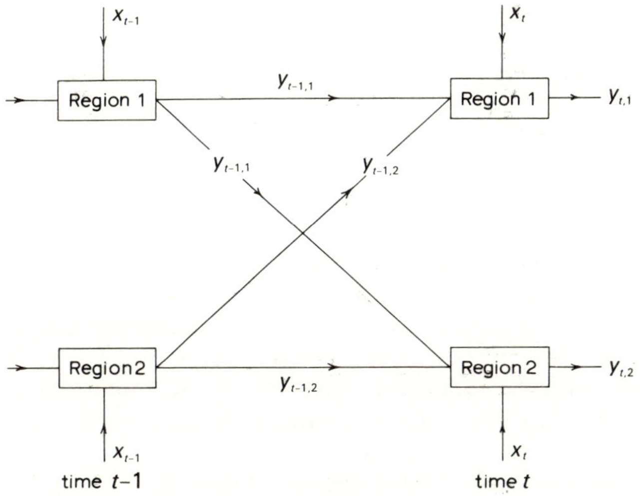

Figure 2.3 A two-region system exhibiting the decoupling of interactions which can be achieved by the reduced form.

variables (one variable for each of the n regions) and $x_t = (x_{t1}, \ldots, x_{tm})'$, is an m-dimensional vector of exogenous variables. The matrices $A_{1t}, \ldots, A_{pt}$, $B_{0t}, \ldots, B_{qt}$ contain the coefficients deriving from the original structural form. The time subscript is used to allow for the possibility of time-varying parameters of the model (see Ch. 5). The subscripts p and q denote the maximum time lag for the endogenous and exogenous variables respectively. Of course Equation (2.8) is a particular case of (2.9), where $p = 1, q = 0, m = 1$ and $n = 2$. The matrices A_{1t} and B_{0t} are given in terms of the original parameters as

$$A_1 = \frac{1}{1 - a_2 b_2} \begin{bmatrix} a_1 + a_2 b_3 & a_3 + a_2 b_1 \\ a_1 b_2 + b_3 & a_3 b_2 + b_1 \end{bmatrix},$$

$$B_0 = \frac{1}{1 - a_2 b_2} \begin{bmatrix} a_4 + a_2 b_4 \\ a_4 b_2 + b_4 \end{bmatrix}$$

where the time subscript t has been omitted for simplicity.

An important observation that can be made from a spatial model in its reduced form, such as Equation (2.9), is that it is similar in form to a one-region system with n variables (Bennett 1979). This means that the mathematical results of multivariable control theory can be directly applied to spatial systems.

The state-space form The structural form and reduced form representations for linear models are familiar to econometricians. However, for the purposes of developing control solutions, it is simpler to use state-space models.

State-space models for linear dynamical systems are usually defined by the following equations:

$$y_t = A_t y_{t-1} + B_t x_t \tag{2.10a}$$

$$w_t = C_t y_t + D_t x_t \tag{2.10b}$$

where x_t denotes inputs, y_t the states and w_t the outputs. Equation (2.10a) is usually termed the state equation and Equation (2.10b) the measurement equation. Equation (2.10b) represents the transformation between the states and the observable outputs of the system. Of course, (2.10b) is irrelevant when the states can be measured directly and in such a case we have $C_t = I$, the identity matrix, $D_t = 0$ the null matrix, and $w_t = y_t$. Frequently, however, $C_t \neq I$, and then the dynamic state vector y cannot be observed or measured directly.

Equation (2.10a), the state equation, is similar to the reduced form representation (2.9). Although the reduced form contains more lagged variables, it will be seen later that it is possible to transform the general reduced form of a linear dynamical system to one which is analogous to the state space form (2.10a). Such system representations are particularly convenient for optimal control strategies and they are used throughout this book. For further details of transfer function and state-space models, see Jacobs (1974) and Aoki (1976).

Throughout this section, the discussion has centred for convenience on deterministic models. The exposition can also be developed for models in which the stochastic error term is included. For example, we have

$$y_t = A_{1t}y_{t-1} + \ldots + A_{pt}y_{tp} + B_{0t}x_t + \ldots + B_{qt}x_{t-q} + e_t \quad (2.11)$$

instead of (2.9), where e_t is a random error term. Such models occur frequently in practice and will be used extensively below.

2.3 *Optimal control solutions: deterministic case*

In the rest of this chapter, a number of examples are worked through to show the way in which control solutions are derived in different situations. The first case concerns systems with structure given by Equation (2.9), i.e. a deterministic structure:

$$y_t = A_{1t}y_{t-1} + \ldots + A_{pt}y_{t-p} + B_{0t}x_t + \ldots + B_{qt}x_{t-q} + \bar{b}_t \quad (2.12)$$

where y_t is a $n \times 1$ vector of system endogenous variables, x_t is a $m \times 1$ vector of system control variables and $\bar{b}_t$ is a $n \times 1$ vector of known exogenous variables which are not subject to control. The matrices $A_{1t}, \ldots, A_{pt}$ are each of dimension $n \times n$ while $B_{0t}, \ldots, B_{qt}$ are each of dimension $n \times m$ and represent sets of known parameters. The A_{it} matrices are autoregressive or feedback terms and the B_{it} matrices are distributed lag terms representing the impact on the n outputs y_t of past values of the control variable x_t. In this case the dimension n could be taken to refer either to the n regions of a given national economy with one endogenous variable per region or to a single region with n endogenous variables.

The form (2.12) is inconvenient for the development of optimal control rules because it is cumbersome and difficult to manipulate. A commonly used method (e.g. Pindyck 1973, Chow 1975, Turnovsky 1977a,b) is to transform the model (2.12) into an equivalent first-order system, which is given by

$$Y_t = A_tY_{t-1} + B_tx_t + b_t \quad (2.13)$$

where

$$
Y_t = \begin{bmatrix} y_t \\ y_{t-1} \\ \vdots \\ y_{t-p+1} \\ \hline x_t \\ x_{t-1} \\ \vdots \\ x_{t-q+1} \end{bmatrix}, \quad B_t = \begin{bmatrix} B_{0t} \\ 0 \\ \vdots \\ 0 \\ \hline I \\ 0 \\ \vdots \\ 0 \end{bmatrix}, \quad b_t = \begin{bmatrix} \bar{b}_t \\ 0 \\ \vdots \\ 0 \\ \hline 0 \\ 0 \\ \vdots \\ 0 \end{bmatrix} \tag{2.14}
$$

$$
A_t = \left[\begin{array}{cccc|cccc} A_{1t} & A_{2t} & \ldots & A_{pt} & B_{1t} & \ldots & & B_{qt} \\ I & 0 & \ldots & 0 & 0 & \ldots & & 0 \\ \vdots & \vdots & & \vdots & \vdots & & & \vdots \\ 0 & 0 & \ldots & I\ 0 & 0 & \ldots & & 0 \\ \hline 0 & 0 & \ldots & 0 & 0 & \ldots & & 0 \\ 0 & 0 & \ldots & 0 & I & \ldots & & 0 \\ \vdots & \vdots & & \vdots & \vdots & & & \vdots \\ 0 & 0 & \ldots & 0 & 0 & \ldots & I & 0 \end{array}\right]
$$

and Y_t is a vector of dimension $(pn + qm) \times 1$. The matrix A_t is termed the companion matrix. There are several reasons why the representation (2.13) is to be preferred. First, it simplifies the algebraic manipulations when the optimal control solution is derived; a first-order dynamic equation is easier to manipulate than one containing higher orders. Secondly, the transformation (2.13) preserves the structure of the original model and allows easier interpretation of the optimal control solutions (Turnovsky 1977a,b). Thirdly, the quadratic control objective function can be developed in a simpler form if it can be expressed in arguments only of the terms of Y_t. Thus, it is not necessary to assume that the weighting matrix for deviations of the controls from desired values is non-singular (Pindyck 1973, Chow 1975). It should be pointed out that the control variable x_t in (2.13) is exactly the same as given in (2.12) while the variable Y_t contains current and possibly lagged endogenous variables as well as current and, possibly, lagged control variables.

After specifying the dynamics of the system, the next essential step in developing an optimal control system is to define the objective function. This measures and weights the deviation of the system variables from the desired, or target, levels. In this book, we shall largely restrict attention to quadratic objective functions such as that given by

$$W = \sum_{t=1}^{T}(Y_t - a_t)'M_t(Y_t - a_t) \tag{2.15}$$

where the vectors $a_1, \ldots, a_T$ are termed target vectors and have the same dimension as Y_t. The matrices $M_1, \ldots, M_T$ are weighting matrices which are assumed to be known, symmetric and positive semi-definite. At time t, the elements of M_t weight the corresponding deviations of elements of Y_t from their target values. Since Y_t contains lagged variables, only the current variables will be relevant at time t. This means that elements of M_t corresponding to lagged variables will take the value zero. In (2.15) we have a fixed-time control problem for T stages. The objective function is written as the sum of squared deviations from $t = 1$ to $t = T$.

The quadratic welfare function has been widely used in spatial control (MacKinnon 1975, Karlqvist *et al.*, 1978, Snickars 1978) and economic control (Theil 1958, Pindyck 1973, Chow 1975, Shupp 1976, Turnovsky 1977a,b) because it avoids cumbersome and complicated mathematical manipulations. However, it has a number of defects. The most important of these is that it assigns the same cost to both positive and negative deviations from targets. Thus, for example, if the target rate of inflation is 6%, an achievement of 5% is treated in the same way as 7%. This defect can be overcome by imposing state inequality constraints (as discussed in Ch. 6) and this disposes of the major objection to the quadratic. However, in the future other more general objective functions should be tried. For example, the piecewise quadratic proposed by Friedman (1973). However, it should be pointed out that doubts have been raised as to whether more complicated objective functions will be any less arbitrary or more determinate than the quadratic (Livesey 1973).

The optimal control problem can be defined either over a finite or over an infinite time horizon. We shall be concerned only with a finite time horizon control problem, i.e. T-stage or fixed-time control.

The optimal control solution Several approaches can be employed to yield the solution of the basic control problem, and that given below uses the method of dynamic programming due originally to Bellman (1957). This method is adopted because it can be applied to the modifications of the control problem which are considered in later chapters. In contrast, the method of Pontryagin *et al.* (1962), based on the minimum principle deriving from the calculus of variations, is suitable only for deterministic control problems and does not in general yield good computational algorithms.

The derivation of the optimal control strategy for the basic problem using dynamic programming given below is based on that given by Chow (1975). First, we restate the basic control problem:

Given the linear model (2.13) *with the initial condition* Y_0, *the optimal control problem is to find the set of control variables* $x_1, \ldots, x_T$ *to minimise the quadratic objective function* (2.15). *The matrices* A_t, B_t *are assumed to be known and* M_t *is a known symmetric positive semi-definite matrix.*

With the method of dynamic programming it is usual to solve the problem 'backwards' step by step, i.e. to solve the problem for the last stage T first, while the problem for the first stage is solved last. For a full discussion of dynamic programming see Dreyfus (1965) and Jacobson & Mayne (1970).

Consider the problem for the last period T. Given all information up to the end of period $T - 1$, the objective function

$$W_T = (Y_T - a_T)'M_T(Y_T - a_T) = Y_T'H_TY_T - 2Y_T'h_T + d_T \tag{2.16}$$

is to be minimised with respect to the controls x_T, and the following definitions are made for ease of generalisation to the multiperiod problem:

$$H_T = M_T, \qquad h_T = M_Ta_T, \qquad d_T = a_T'M_Ta_T \tag{2.17}$$

Substituting (2.13) for Y_T we can rewrite W_T as

$$W_T = (A_TY_{T-1} + B_Tx_T + b_T)'H_T(A_TY_{T-1} + B_Tx_T + b_T) - 2(A_TY_{T-1} + B_Tx_T + b_T)' + d_T \tag{2.18}$$

It is well known from calculus that we need to find the derivative of W_T with respect to x_T and set it to zero in order to get the optimal x_T yielding minimum W_T. That is,

$$\frac{\partial W_T}{\partial x_T} = 2B_T'H_T(A_TY_{T-1} + B_Tx_T + b_T) - 2B_Th_T = 0 \tag{2.19}$$

and on solving for x_T the optimal control for the last period is given by

$$x_T'^{*} = G_T Y_{T-1} + g_T \tag{2.20}$$

where

$$G_T = -(B_T' H_T B_T)^{-1} B_T H_T A_T \tag{2.21}$$

$$g_T = (B_T' H_T B_T)^{-1} B_T' (h_T - H_T b_T) \tag{2.22}$$

The minimum of the objective function for the last period is easily found if (2.20) is substituted for x_T in (2.18), giving

$$\begin{aligned} \min W_T = W_T^* = {} & Y_{T-1}'(A_T + B_T G_T)' H_T (A_T + B_T G_T) Y_{T-1} \\ & + 2Y_{T-1}'(A_T + B_T G_T)'(H_T B_T g_T + H_T b_T - h_T) \\ & + (B_T g_T + b_T)' H_T (B_T g_T + b_T) \\ & - 2(B_T g_T + b_T)' h_T + d_T \end{aligned} \tag{2.23}$$

which is a quadratic function of Y_{T-1}. Equations (2.20)–(2.22) give the optimal control setting of x at time T as a linear function of Y_{T-1}.

Next, we need to consider the problem from period $T - 1$ onwards. From the principle of optimality in dynamic programming (Bellman's principle) we need only to minimise

$$\begin{aligned} W_{T-1} = {} & Y_{T-1}' M_{T-1} Y_{T-1} - 2Y_{T-1}' M_{T-1} a_{T-1} \\ & + a_{T-1}' M_{T-1} a_{T-1} + W_T^* \end{aligned} \tag{2.24}$$

with respect to x_{T-1}. The objective function at period $T - 1$ given by (2.24) is made up of two parts: the contribution from the present stage $T - 1$ as well as the minimum cost W_T^* from period T onwards. Replacing W_T^* in (2.24) by (2.23) we have

$$W_{T-1} = Y_{T-1}' H_{T-1} Y_{T-1} - 2Y_{T-1}' h_{T-1} + d_{T-1} \tag{2.25}$$

where

$$\begin{aligned} H_{T-1} & = M_{T-1} + (A_T + B_T G_T)' H_T (A_T + B_T G_T) \\ & = M_{T-1} + A_T' H_T (A_T + B_T G_T) \end{aligned} \tag{2.26}$$

$$h_{T-1} = M_{T-1}a_{T-1} + (A_T + B_T G_T)'(h_T - H_T B_T g_T - H_T b_T) \quad (2.27)$$
$$= M_{T-1}a_{T-1} + A'_T(h_T - H_T B_T g_T - H_T b_T)$$

$$d_{T-1} = a'_{T-1}M_{T-1}a_{T-1} + (B_T g_T + b_T)'H_T(B_T g_T + b_T) - 2(B_T g_T + b_T)'h_T + d_T \quad (2.28)$$

Since the function to be minimised (Eq. 2.25) is of the same form as the expression (2.16), we can repeat the process from (2.16) to (2.28) and replace the time subscript $T - 1$ for T, and so on. Thus the solution to the T-stage control problem is complete. The optimal control policy at each period is given by (2.20), i.e.

$$x_t^* = G_t Y_{t-1} + g_t, \qquad t = 1, 2, \ldots, T \quad (2.29)$$

To compute the values of the optimal controls at each period, the quantities $G_1, \ldots, G_T$ and $g_1, \ldots, g_T$ are needed. They can be easily evaluated as follows:

(a) With the initial condition $H_T = M_T$, solve the set of equations (2.21) and (2.26) alternately backward in time from $t = T$ to $t = 1$ yielding the matrices $G_T, H_{T-1}, G_{T-1}, H_{T-2}, \ldots, G_1$.
(b) Next, with the initial condition $h_T = M_T a_T$, solve the equations (2.22) and (2.27) alternately backward in time from $t = T$ to $t = 1$ to give the vectors $g_T, h_{T-1}, g_{T-1}, h_{T-2}, \ldots, g_1$.
(c) Finally, with G_t and g_t computed for $t = 1, \ldots, T$, the optimal control x_t^* is determined by the linear control rule (2.29) dependent on Y_{t-1}.

Together with the optimal controls, the minimum value of the objective function (2.15) is given by (2.23) with the time subscript T replaced by 1 and given the initial condition Y_0. Of course, the scalar quantity d is obtained by solving (2.28) backwards in time given the initial condition $d_T = a'_T M_T a_T$.

For this deterministic control problem, it is possible to calculate in period 1 the optimal controls x_t for $t = 1, \ldots, T$ according to (2.29). Furthermore, these optimal control policies are optimal linear feedback (closed-loop) policies, i.e. the control policies are functions of the states and hence include feedback from the output (Lee 1964, Shinners 1973). Another feature of the optimal control rule (2.29) is that determination of the current optimal policy requires consideration of the future policies. This is due to the multistage nature of the control problem.

The matrices H_t are calculated recursively according to the difference equation (2.26) with the initial condition $H_T = M_T$. This equation is also known as the matrix Riccati equation. It may be easily checked that the matrices H_t are symmetric. This follows from the assumption that M_T, $M_{T-1}, \ldots, M_1$ are all symmetric matrices.

An example The following is an introductory example which exhibits the character and properties of the control solution in the deterministic case.

Given the linear model

$$Y_t = \begin{bmatrix} 3 & 0 & -1 \\ -1 & 4 & 1 \\ 1 & 1 & 1 \end{bmatrix} Y_{t-1} + \begin{bmatrix} 2 & 1 \\ 1 & 1 \\ 1 & 1 \end{bmatrix} X_t + \begin{bmatrix} 10 \\ 20 \\ 15 \end{bmatrix}$$

with initial condition $Y_0 = (21, 21\ 7)'$, minimise the two-stage objective function

$$W = \sum_{t=1}^{2} (Y_t - a_t)'(Y_t - a_t)$$

with respect to the controls x_1 and x_2. The targets are set at $a_1 = (100, 100, 50)'$ and $a_2 = (150, 150, 150)'$.

Computing backwards in time from $t = 2$ to $t = 1$ we obtain, from Equations (2.17), (2.21) and (2.26),

$$H_2 = \begin{bmatrix} 1 & 0 & 0 \\ 0 & 1 & 0 \\ 0 & 0 & 1 \end{bmatrix}, \quad G_2 = \begin{bmatrix} -3 & 5/2 & 2 \\ 3 & -5 & -3 \end{bmatrix}$$

$$H_1 = \begin{bmatrix} 3 & -3 & 0 \\ -3 & 11/2 & 0 \\ 0 & 0 & 1 \end{bmatrix}, \quad G_1 = \begin{bmatrix} -8 & 8 & 4 \\ 62/7 & -100/7 & -6 \end{bmatrix}$$

Similarly, from equations (2.17), (2.22) and (2.27), the following vectors are calculated:

$$h_2 = \begin{bmatrix} 150 \\ 150 \\ 150 \end{bmatrix}, \quad g_2 = \begin{bmatrix} 15/2 \\ 125 \end{bmatrix}, \quad h_1 = \begin{bmatrix} 105 \\ 185/2 \\ 50 \end{bmatrix}, \quad g_1 = \begin{bmatrix} 90 \\ 100/7 \end{bmatrix}$$

The optimal control settings are then given by (2.29), i.e.

$$x_1^* = G_1 Y_0 + g_1 = \begin{bmatrix} 118 \\ -992/7 \end{bmatrix},$$

$$x_2^* = G_2Y_1 + g_2 = \begin{bmatrix} -3179/14 \\ 11430/14 \end{bmatrix}$$

where Y_1 is calculated by substituting x_1^* into the linear model.

2.4 *Optimal control solutions: stochastic case*

We consider next the basic control problem subject to a stochastic linear dynamic model given by

$$Y_t = A_tY_{t-1} + B_tx_t + b_t + u_t \tag{2.30}$$

This is identical to Equation (2.13) except that a random disturbance term u_t has been added. Initially it is assumed that u_t is a vector of random variables of dimension $(pn + qm) \times 1$ with zero mean and co-variance matrix given by

$$\text{Cov}\,(u_t, u_s) = \begin{cases} R_u & \text{if } t = s \\ 0 & \text{if } t \neq s \end{cases}$$

and that u_t is independent of $Y_{t-1}, Y_{t-2}, \ldots, x_t, x_{t-1}, \ldots$ The deterministic objective function (2.15) will no longer be appropriate here. It has to be modified to take into account the uncertainty present in the system dynamics. The simplest and most usual way to undertake this modification is to use the expected value of the objective function, namely

$$E_0(W) = E_0\left[\sum_{t=1}^{T} (Y_t - a_t)'M_t(Y_t - a_t)\right] \tag{2.31}$$

where E_0 denotes the expectation conditional on the initial condition Y_0. In general, E_t represents the conditional expectation given all information up to the end of period t. The stochastic control problem can then by stated as follows:

> *Given the linear model* (2.30) *with the initial condition* Y_0, *the stochastic control problem is to determine the set of control variables* $x_1, \ldots, x_T$ *to minimise the objective function* (2.31). *The matrices* A_t, B_t *are assumed to be known and* M_t *is a known symmetric positive semi-definite matrix.*

As before, and following Chow (1975), the method of dynamic programming is used to solve the stochastic control problem. Considering the last period T and with all available information given up to period $T-1$, the function

$$\begin{aligned} V_T &= E_{T-1}(Y_T - a_T)'M_T(Y_T - a_T) \\ &= E_{T-1}(Y'_T H_T Y_T - 2Y'_T h_T + d_T) \end{aligned} \tag{2.32}$$

is to be minimised with respect to x_T. H_T, h_T and d_T are defined according to (2.17). Substituting Y_T in (2.32) by (2.30) and taking expectations we have that

$$\begin{aligned} V_T = &(A_T Y_{T-1} + B_T x_T + b_T)'H_T(A_T Y_{T-1} + B_T x_T + b_T) \\ &- 2(A_T Y_{T-1} + B_T x_T + b_T)'h_T \\ &+ E_{T-1}(u'_{T-1} H_T u_T) + d_T \end{aligned} \tag{2.33}$$

To obtain x_T^*, the optimal control for period T, we set the derivative of V_T with respect to x_T to zero and then solve for x_T^*. That is,

$$\frac{\partial V_T}{\partial x_T} = 2B'_T H_T(A_T Y_{T-1} + B_T x_T + b_T) - 2B'_T h_T = 0 \tag{2.34}$$

so that, on solving for x_T,

$$x_T^* = G_T Y_{T-1} + g_T. \tag{2.35}$$

where G_T and g_T are given by (2.21) and (2.22) respectively. Replacing x_T in (2.33) by (2.35) gives the minimum welfare cost for period T; that is

$$\begin{aligned} \min V_T = V_T^* = &Y'_{T-1}(A_T + B_T G_T)'H_T(A_T + B_T G_T)Y_{T-1} \\ &+ 2Y'_{T-1}(A_T + B_T G_T)'(H_T B_T g_T + H_T b_T - h_T) \\ &+ (B_T g_T + b_T)'H_T(B_T g_T + b_T) \\ &- 2(B_T g_T + b_T)'h_T \\ &+ d_T + E_{T-1}(u'_T H_T u_T) \end{aligned} \tag{2.36}$$

which is quadratic in Y_{T-1}.

From Bellman's principle of optimality, we need to minimise the function

$$\begin{aligned} V_{T-1} = E_{T-2}(&Y'_{T-1} M_{T-1} Y_{T-1} - 2Y'_{T-1} M_{T-1} a_{T-1} \\ &+ a'_{T-1} M_{T-1} a_{T-1} + V_T^*) \end{aligned} \tag{2.37}$$

with respect to x_{T-1} when the problem is considered from period $T-1$ onwards.

Replacing V_T^* in (2.37) by (2.36) gives

$$V_{T-1} = E_{T-2}(Y'_{T-1}H_{T-1}Y_{T-1} - 2Y'_{T-1}h_{T-1} + d_{T-1}) \tag{2.38}$$

where H_{T-1} and h_{T-1} are given as before by (2.26) and (2.27) respectively, while

$$\begin{aligned} d_{T-1} = a'_{T-1}M_{T-1}a_{T-1} &+ (B_Tg_T + b_T)'H_T(B_Tg_T + b_T) \\ &- 2(B_Tg_T + b_T)'h_T \\ &+ d_T + E_{T-1}(u'_TH_Tu_T) \end{aligned} \tag{2.39}$$

The function V_{T-1} to be minimised with respect to x_{T-1} has the same form as the expression (2.32) but with the subscript T replaced by $T - 1$. Repeating the optimisation procedure, the optimal control policy at period t is thus given by

$$x_t^* = G_tY_{t-1} + g_t \quad \text{for} \quad t = 1, \ldots, T. \tag{2.40}$$

and the associated minimum value of the objective function is

$$\begin{aligned} \min E_0(W) = Y'_0(A_l + B_1G_1)'H_1(A_1 + B_1G_1)Y_0 \\ + 2Y'_0(A_1 + B_1G_1)'(H_1B_1g_1 + H_1b_1 - h_1) \\ + (B_1g_1 + b_1)'H_1(B_1g_1 + b_1) \\ - 2(B_1g_1 + b_1)'h_1 \\ + d_1 + E_0(u_1H_1u_1) \end{aligned} \tag{2.41}$$

where d_1 is obtained by solving (2.39) backward in time given the initial condition $d_T = a'_TM_Ta_T$.

A striking feature of the optimal control solution (2.40) for the stochastic case is its similarity with that of the deterministic case; both control rules (2.29) and (2.40) have identical linear feedback equations. This arises from the principle of *certainty equivalence* (Simon 1956, Theil 1957, Åstrom 1970). This principle allows us to reduce the stochastic control problem to one that is deterministic by replacing the random variables involved by their expected values. Solving this deterministic problem results in a solution which is called the certainty equivalent solution for the stochastic control problem. If the random variables involved are only the error terms u_t, then an important result on applying the certainty equivalence principle is that the optimal solution of the stochastic control problem is the same as for the deterministic problem. In other words, the certainty equivalent solution is optimal for the deterministic *and* stochastic problem. This property is specific to optimal control problems with a linear dynamic model and a quadratic objective function. (Despite this property, however, there are situations where the certainty equivalent solution is not optimal; these are described in later chapters.)

Even though the optimal control equations are the same for both deterministic and stochastic cases, there is a basic difference. In the deterministic case it is possible to determine numerically in period 1 the time path of the optimal x_t^* for all t, but in the stochastic case the optimal x_t^* can be calculated only after Y_{t-1} is observed. This is due to the presence of the random disturbances u_t which make the future values of the states unknown at time $t = 1$.

Because of the close similarities of the stochastic and deterministic cases a worked example is not necessary here. However, the control problem discussed here has found numerous applications especially in the study of policies for economic stabilisation. A good example is provided by Pindyck (1973), who used a linear dynamic model of the US economy with 28 state variables and three control variables (tax surcharge, government expenditures and money supply).

2.5 *Control with incomplete state information*

So far it has been assumed that the outputs and states of the system are identical, i.e. that $C = I$ and $D = 0$ in Equation (2.10b). This will not be true in many situations. For cases in which the states are no longer directly measurable and the addition of a measurement operation to the state space form is required, i.e. the states can only be measured indirectly through an $s \times 1$ output vector z_t. The system model must be written as

$$Y_t = A_t Y_{t-1} + B_t x_t + b_t + u_t \tag{2.42a}$$

$$z_t = C_t Y_t + v_t \tag{2.42b}$$

Equation (2.42a) is identical to (2.30) with the same assumptions on the error term u_t. However, the dimension of z_t need not be equal to that of Y_t. The matrix C_t is a known $s \times (pn + qm)$ matrix of parameters and v_t is a random vector sequence with zero mean and covariance matrix

$$\text{Cov}(v_t, v_s) = \begin{cases} R_v & \text{if } t = s \\ 0 & \text{if } t \neq s \end{cases}$$

which is assumed to be known and non-singular.

Although it is more common to encounter indirectly measurable states in engineering applications of optimal control theory, the developments discussed in this section are applicable to the control of unknown or partially known spatial systems, in particular the case where the unknown model parameters are time-varying. This case is considered in more detail in Chapter 5. For the moment, we will give the optimal control solution to the problem with incomplete state information.

In the previous developments, it has been implicitly assumed that the decision taken at time t is made with knowledge of the past states and controls up to time $t - 1$. It is now assumed that $Y_0, Y_1, \ldots, Y_{t-1}$ cannot be observed and the decision made at time t is based on knowledge of $x_1, \ldots, x_{t-1}$ and of $z_0, \ldots, z_{t-1}$. With these assumptions the control problem can be stated as follows:

Given the model with indirectly measurable states (2.42), *determine the sequence of control variables $x_1, \ldots, x_T$ to minimise the quadratic welfare function* (2.31). *The matrices A_t, B_t are assumed known and M_t is a known symmetric positive semi-definite matrix.*

Our approach to this problem again uses the method of dynamic programming. It turns out that the optimal control rule is a linear function of the estimate of the state, namely its conditional mean. The technique of Kalman filtering (Kalman 1960, Chow 1975) allows this estimation process to be undertaken in a recursive manner (this also is useful, as will be seen in Chapter 5, in situations where the parameters of the model are unknown and time-varying).

To obtain the solution consider the control problem with incomplete state information at the last period T. It is required to minimise with respect to the controls x_T the objective function (2.32), which on expansion may be expressed as

$$\begin{aligned} V_T &= E_{T-1}(Y_T'H_TY_T - 2Y_T'h_T + d_T) \\ &= E_{T-1}(Y_{T-1}'A_T'H_TA_TY_{T-1}) \\ &\quad + 2(B_Tx_T + b_T)'H_TA_T(E_{T-1}Y_{T-1}) \\ &\quad + (B_Tx_T + b_T)'H_T(B_Tx_T + b_T) \\ &\quad - 2(E_{T-1}Y_{T-1}')A_T'h_T \\ &\quad - 2(B_Tx_T + b_T)'h_T + d_T + E_{T-1}(u_T'H_Tu_T) \end{aligned} \tag{2.43}$$

The difference between (2.43) and (2.33) is that, in the former, Y_{T-1} cannot be treated as given; it is only 'known' via Equation (2.42b). Of course, in (2.43) the assumption that u_t is independent of Y_{t-1}, $Y_{t-2}, \ldots,$ is required. Proceeding as before, i.e. solving the equation

$\partial V_{T-1}/\partial x_T = 0$ for x_T, it is easily checked that the optimal control at time T, x_T^*, is given by

$$x_T^* = G_T(E_{T-1}Y_{T-1}) + g_T \tag{2.44}$$

where G_T and g_T are the same as given in (2.21) and (2.22) respectively. Notice that this optimal control policy for period T is the same linear feedback equation given in Section 2.4. The only difference is that in (2.44) the conditional mean of Y_{T-1} is used instead of Y_{T-1}. For the moment write $\bar{Y}_{T-1}$ to denote $E_{T-1}Y_{T-1}$ and assume that it can be evaluated. Substituting (2.44) for x_T in (2.43) gives the minimum of the objective function at time T:

$$\begin{aligned}\min V_T = V_T^* &= E_{T-1}[Y'_{T-1}(A_T + B_TG_T)'H_T(A_T + B_TG_T)Y_{T-1}] \\ &+ 2\bar{Y}'_{T-1}(A_T + B_TG_T)'(H_TB_TG_T + H_Tb_T - h_T) \\ &+ (B_TG_T + b_T)'H_T(B_TG_T + b_T) \\ &- 2(B_TG_T + b_T)'h_T + d_T \\ &+ E_{T-1}(u'_TH_Tu_T) \\ &+ E_{T-1}[(\bar{Y}_{T-1} - Y_{T-1})'G'_TB'_TH_TB_TG_T(\bar{Y}_{T-1} - Y_{T-1})]\end{aligned} \tag{2.45}$$

Next, we have to consider the problem from period $T - 1$ onwards. By the principle of optimality, we need only to minimise the objective function

$$\begin{aligned}V_{T-1} = E_{T-2}(&Y'_{T-1}M_{T-1}Y_{T-1} - 2Y'_{T-1}M_{T-1}a_{T-1} \\ &+ a'_{T-1}M_{T-1}a_{T-1} + V_T^*)\end{aligned} \tag{2.46}$$

with respect to x_{T-1}. Using the identity, given by Chow (1975), that

$$E_{T-2}[E_{T-1}f(Y)] = E_{T-2}f(Y) \tag{2.47}$$

and replacing V_T^* in (2.46) by (2.45), it may be shown that

$$V_{T-1} = E_{T-2}(Y'_{T-1}H_{T-1}Y_{T-1} - 2Y'_{T-1}h_{T-1} + d_{T-1}) \tag{2.48}$$

where H_{T-1} and h_{T-1} are the same as given in (2.26) and (2.27) respectively, while d_{T-1} satisfies

$$\begin{aligned} d_{T-1} = {} & a'_{T-1}M_{T-1}a_{T-1} + (B_{T}g_{T} + b_{T})'H_{T}(B_{T}g_{T} + b_{T}) \\ & - 2(B_{T}g_{T} + b_{T})'h_{T} + d_{T} \\ & + E_{T-1}(u'_{T}H_{T}u_{T}) \\ & + E_{T-1}[(\bar{Y}_{T-1} - Y_{T-1})'G'_{T}B'_{T}H_{T}B_{T}G_{T}(\bar{Y}_{T-1} - Y_{T-1})] \end{aligned} \tag{2.49}$$

Observe that (2.48) is identical to (2.43) but with $T - 1$ replacing T. This implies that the T-period control problem is now solved. Hence the optimal control policy at time t takes the form

$$x_t^* = G_t\bar{Y}_{t-1} + g_t, \qquad t = 1, \ldots, T \tag{2.50}$$

which is the familiar linear feedback control rule on $\bar{Y}_{T-1}$. The associated minimum value of the T-period objective function is given by (2.45) with the subscript T replaced by 1 and d_1 is obtained by solving (2.49) backward in time given the initial condition $d_T = a'_T M_T a_T$.

The optimal control rule (2.50) results in a *separation theorem* (Vishwakarma 1974, Chow 1975, Aoki 1976). Following Chow, the separation theorem may be stated as follows:

For linear dynamical systems with known parameters and quadratic objective functions, the separation theorem states that when the states of the system are not directly measurable the optimal control solution separates into two parts. First, the optimal control x_t^* is given as a linear function of the conditional mean of Y_{t-1}. These optimal equations are identical to those when the system states are directly measurable. The second part considers the problem of evaluating $\bar{Y}_{t-1}$.

Note that for the first part of the solution the certainty equivalent solution with the uncertain Y_{t-1} is replaced by its expected value and it is optimal for the problem with incomplete state information. In other words, the principle of certainty equivalence applies.

In order to find the setting of the optimal control x_t^* by the linear feedback rule (2.50), it is necessary to determine $\bar{Y}_{t-1}$. In addition, if the optimal value of the objective function V_1^* is needed, then it is also necessary to evaluate the covariance matrix $E_{t-1}(Y_{t-1} - \bar{Y}_{t-1})(Y_{t-1} - \bar{Y}_{t-1})'$. When the random disturbances u_t and v_t satisfy certain distributional conditions, the conditional mean and its associated covariance matrix can be determined recursively. This result is due to Kalman (1960). It is this method which has since been referred to as the Kalman filter.

To determine the Kalman filter it is necessary to assume, in addition to the assumptions on u_t and v_t made above, that the random disturbance variables are both normally distributed and that $\mathrm{Cov}(u_t, v_t) = 0$. Furthermore, the initial stage Y_0 is assumed to be independent of u_t and v_t and is normally distributed with known mean $\bar{Y}_0$ and covariance Γ_0. It will be convenient to introduce the following notation:

$$E_{t-1}(Y_t) = E(Y_t/z_{t-1}) = Y_{t/t-1}$$

$$E(Y_t - Y_{t/t-1})(Y_t - Y_{t/t-1})' = \Gamma_{t/t-1}$$

so that

$$\bar{Y}_t = E(Y_t/z_t) = Y_{t/t}$$

The Kalman filter algorithm is then given (Kalman 1960, Lee 1964, Duncan & Horn 1972, Chow 1975, Bennett 1979, Ch. 5) as

$$\begin{aligned} Y_{t/t} = {} & (A_t Y_{t-1/t-1} + B_t x_t + b_t) \\ & + \Phi_t[z_t - C_t(A_t Y_{t-1/t-1} + B_t x_t + b_t)] \end{aligned} \tag{2.51a}$$

$$\Phi_t = \Gamma_{t/t-1} C_t'[C_t \Gamma_{t/t-1} C_t' + R_v]^{-1} \tag{2.51b}$$

$$\Gamma_{t/t-1} = A_t \Gamma_{t-1/t-1} A_t' + R_u \tag{2.51c}$$

$$\Gamma_{t/t} = \Gamma_{t/t-1} - \Phi_t C_t \Gamma_{t/t-1} \tag{2.51d}$$

$$\Gamma_{0/0} = \Gamma_0 \tag{2.51e}$$

$$Y_{0/0} = \bar{Y}_0 \tag{2.51f}$$

The matrix Φ is sometimes called the Kalman gain matrix. It is given by (2.51b) so that the Kalman filter gives a minimum variance unbiased estimator of Y_t.

2.6 Conclusion

This chapter has been concerned with translating spatial system models into the form that is most suitable for optimal control analysis. For the various cases of optimal control considered, details have been provided for deriving the optimal solutions. However, some general remarks on the solutions as well as the control problems may be useful.

First, the optimal control equations for the basic control problems considered so far assume that the inverse of the matrix $B_t'H_tB_t$ exists, i.e. that the matrix is non-singular. When the matrix is singular, it has been suggested that the pseudo-inverse (Rao 1973) can be used instead (Chow 1972, Aoki 1976). The necessary and sufficient conditions for the matrix $B_t'H_tB_t$ to be non-singular have been obtained by Garbade (1976), who also considers the problem of the existence and uniqueness of a solution when $B_t'H_tB_t$ is singular. As pointed out by Garbade, the conditions given can hardly be verified before the computations. In practice then, one discovers that $B_t'H_tB_t$ is singular in the actual course of the recursive calculations that have to be performed to get the optimal control settings. When $p = 1$, $q = 0$ in (2.9) or (2.12) and if $m > n$ (i.e. there are more control variables than states) then the matrix $B_t'H_tB$ is always singular. However, when the state vector is augmented with the control vector and when the number of target variables is not less than the number of controls, then the formulation of the control problems in the previous sections ensures that $B_t'H_tB_t$ is non-singular (Chow 1975) and hence yields a unique solution. These assumptions basically assure that the system is both controllable and observable. A good discussion of the practical aspects of these conditions is given by Blechman *et al*. (1975).

Other simplifications made in the control problems considered so far are that there are no constraints on state or controls and that all the matrices of parameters are known with certainty. These restrictions can be relaxed, but only at the expense of more complicated and sometimes sub-optimal control solutions. The case of constrained control solutions is considered in the next chapter, and the cases with unknown model parameters, together with constraints on the control variables, will be considered in Chapters 4 and 5.

3 *Spatial control with constraints on control variables*

3.1 Optimal control under constraints

In this chapter, the basic spatial control problem will be modified to take account of situations in which there are constraints on the control variables. Such constrained control problems arise in many situations, but their solution requires the development of a new set of control solutions. A major example of such constrained control is the allocation of cash grants to lower level governments, subject to a total sum of money being available, no government being given a negative allocation, and subject to the dynamic properties of the system. This application is discussed further in Chapter 7. Other applications concern control of environmental pollution where the state variables (levels of pollution and productive capital) must satisfy fixed terminal conditions (Nijkamp 1976, 1980), or in national settlement system planning in which population growth due to migration is restricted to certain areas (MacKinnon 1975).

In general, constraints can take any mathematical form. However, in this book, linear equality and inequality constraints are the centre of concern and these arise naturally from the analysis of important geographical problems from the standpoint of optimal control theory. The presence of inequality constraints on functions of state and/or control variables imposes a major difficulty in obtaining a global optimum in control solutions. Various methods for obtaining a solution have been suggested (Jacobs 1974, Chan & Maille 1975) for saturation constraints on the controllers, i.e. for cases in which the control variable is restricted to lie within a specified interval. One of these methods, based on the minimum principle of Pontryagin, is described by Jacobs (1974), who also shows that the control law for deterministic control problems with a quadratic objective function can be given by a bang-bang control provided that the control variables do not appear explicitly in the objective function. This bang-bang control permits each control variable to take only one of two bounding values specified by the saturation constraint. An alternative method, proposed by Chan and Maille (1975), requires the controllers to appear explicitly in the objective function. Their method is iterative and is based on the gradient method of mathematical programming (see Hadley 1964).

Table 3.1 Typology of methods for solution of constrained control problems.

Study	*Time domain*	*Objective functional*	*Dynamic model*	*Terminal constraints*	*Constraints on control variable*	*Solution method*
Evtushenko and MacKinnon (1976)	discrete	quadratic tracking problem in the state variables (also an exponential)	linear in state and controls, with exogenous term	none	budget constraint, non-negativity constraints, linear equalities (accounting relation)	numerical, modified penalty function
Mehra (1975)		quadratic tracking problem in the state variables	time-invariant linear in state and controls	none	budget constraint, non-negativity constraints, linear equalities	numerical, branch and bound integer programming
Tan (1979a)		quadratic tracking problem in the state variables	linear in state and controls	none	linear equalities	numerical, dynamic programming
Tan (1980)		quadratic tracking problem in the state variables	linear in state and controls	none	linear equalities, linear inequalities on controls, non-negativity constraints	numerical, dynamic programming, and then linear programming using the dual simplex method
Ho and Brentani (1963)		quadratic in terminal state	in practice, linear in state and controls	yes	bounds on absolute value of controls	numerical, non-linear programming using generalised gradient projection
Ohno (1978)		general	general	yes	inequalities/ equalities on state and controls	numerical, differential dynamic programming algorithm
Propoi and		linear in state and	linear in state and	yes	budget constraints,	numerical, linear

				constraint)		conditions
Barber and Ralston (1980)	continuous	non-linear in state variable	linear in the control	no	non-negativity and upper bound	analytic (Pontryagin), non-linear feedback with bounds on the control
Clark *et al*. (1979)		cross product in state and controls, and linear in the controls	non-linear in state, and cross product in state and control	no	non-negativity and upper bound	analytic (differential inclusions), switching behaviour
Fujita (1976)		linear in state and controls	linear in controls	no	non-negativity	analytic (Pontryagin), general optimality conditions
Martin and Jacobson (1979)		quadratic in controls, with quadratic terminal state value (required)	linear in state and controls	yes (special case)	controls are elements of a closed cone	analytic (Pontryagin), optimal controls in feedback form are linear-in-half-spaces
Nicoletti *et al*. (1975)		quadratic in state variables and linear in terminal state	cross products in state and controls and linear in the states	no	budget inequality, non-negativity and upper bound	analytic (Pontryagin), phase plane analysis
Ralston (1979)		linear in the state, and non-linear in the control	cross product in state and control and linear in the state	no	non-negativity and upper bound	analytic (Pontryagin), non-linear feedback with bounds on the control
Lee (1982)		quadratic in controls	exogenous term	yes	upper and lower bounds on controls	analytic (Pontryagin), optimal controls constant over time if constraint non-binding

Source: Lee (1982).

Recently Lee (1982) has provided a typology of possible approaches to constrained control problems in terms of whether they are discrete or continuous, the form of their objective function, the nature of the dynamic model of the state equation, and the form of the constraints. This typology, which is reproduced in Table 3.1, shows both the complexity and the richness of the solution strategies that have been developed. It also shows the location of the solution methods developed in this book relative to other forms which are available. It can generally be concluded that the solutions which are developed below represent a very general approach, but are limited to discrete systems.

The various techniques for the solution of control problems with a quadratic objective function and with saturation constraints on the control variables do not lend themselves to simple modifications which allow solutions when there are linear equality and non-negativity constraints on the controllers. For example, the bang-bang control solution need not necessarily satisfy the additional linear equality constraints imposed on the controllers. Hence, new methods of solution are required. This chapter describes one of them. It is based on the method of dynamic programming which is combined with an adjustment mechanism which alters the solution to satisfy both types of constraint. This solution is given in Section 3.3, while Section 3.2 deals with the case in which there are only linear equality constraints on the controllers. It turns out that the methods proposed to solve the control problems given in Sections 3.2 and 3.3 can be modified to deal with the case where the linear dynamic model is stochastic, i.e. where an additive error is present in the state equation as in Equation (2.30). This feature has been absent from the previous numerical techniques for solving optimal control problems with inequality constraints on control variables.

3.2 *Optimal control with linear equality constraints*

For convenience, the solution to the optimal control problem with linear equality constraints on the control variables is first derived for the deterministic case. These results were first given by Tan (1979a,b). The modified deterministic control problem may be stated as follows:

Given the linear dynamic model (2.13) *with the initial condition* Y_0, *the equality-constrained optimal control problem is to find the set of control variables* $x_1, \ldots, x_T$ *to minimise the quadratic objective function* (2.15) *when the control variables* x_t *are subject to a set of* k *independent linear constraints*

$$F_t x_t = f_t, \qquad t = 1, \ldots, T \tag{3.1}$$

where k is less than m. In (3.1), *for each t, f_t is a $k \times 1$ vector of known quantities and F_t is a $k \times m$ matrix of known constants.*

The only difference between this control problem and the basic deterministic control problem of Chapter 2 is the addition of the constraints (3.1). The solution to this modified control problem can be derived using the method of dynamic programming, and this is the approach followed below. As a result, the solution has much in common with that developed in Chapter 2.

As before, we consider the problem for the last period T. Given all information available up to the end of period $T - 1$, we minimise, with respect to the controls x_T, the objective function given by

$$\begin{aligned} W_T &= (Y_T - a_T)'M_T(Y_T - a_T) \\ &= Y'_TH_TY_T - 2Y'_Th_T + d_T \end{aligned} \tag{3.2}$$

By using the model (2.13) for Y_T, this can be rewritten as

$$\begin{aligned} W_T = {} & (A_TY_{T-1} + B_Tx_T + b_T)'H_T(A_TY_{T-1} + B_Tx_T + b_T) \\ & - 2(A_TY_{T-1} + B_Tx_T + b_T)'h_T + d_T \end{aligned} \tag{3.3}$$

The initial quantities H_T, h_T and d_T are as defined by (2.17). The minimisation of (3.3) must be carried out subject to the constraint $F_Tx_T = f_T$. To solve this, form the Lagrangian

$$L_T = W_T - 2\lambda'_T(f_T - F_Tx_T) \tag{3.4}$$

where λ_T is a $k \times 1$ vector of Lagrange multipliers. As the objective function is quadratic and convex, the necessary and sufficient conditions for an optimum are given by

$$\begin{aligned} \frac{\partial L_T}{\partial x_T} = {} & 2B'_TH_T(A_TY_{T-1} + B_Tx_T + b_T) \\ & - 2B'_Th_T + 2F'_T\lambda_T = 0 \end{aligned} \tag{3.5}$$

$$\frac{\partial L_T}{\partial \lambda_T} = -2(f_T - F_Tx_T) = 0 \tag{3.6}$$

which are the basic results from the Kuhn–Tucker theory for non-linear programming (see, for example, Pindyck 1973). Define

$$D_t = B'_tH_tB_t \quad \text{and} \quad P_t = F_tD_t^{-1}F'_t \quad \text{for} \quad t = 1, \ldots, T \tag{3.7}$$

where it is assumed that each expression is invertible. Then it follows from (3.5) that

$$x_T = D_T^{-1}(-B'_T H_T A_T Y_{T-1} + B'_T h_T - B'_T H_T b_T - F'_T \lambda_T) \quad (3.8)$$

Pre-multiplying both sides of (3.8) by F_T and using (3.6), we have, after some simplification,

$$\lambda_T = P_T^{-1}(F_T G_T Y_{T-1} + F_T g_T - f_T) \quad (3.9)$$

where G_T and g_T are as given by (2.21) and (2.22) respectively. Substituting (3.9) into (3.8) we obtain the optimal control solution for the last period:

$$x_T^* = \hat{G}_T Y_{T-1} + \hat{g}_T \quad (3.10)$$

where

$$\hat{G}_T = G_T - D_T^{-1} F'_T P_T^{-1} F_T G_T \quad (3.11)$$

$$\hat{g}_T = g_T - D_T^{-1} F'_T P_T^{-1} F_T g_T + D_T^{-1} F'_T P_T^{-1} f_T \quad (3.12)$$

The minimum value of the objective function for the last period is easily found if we substitute (3.10) for x_T into (3.3) giving

$$\begin{aligned} \min W_T = W_T^* &= Y'_{T-1}(A_T + B_T\hat{G}_T)' H_T (A_T + B_T \hat{G}_T) Y_{T-1} \\ &+ 2Y'_{T-1}(A_T + B_T\hat{G}_T)'(H_T B_T \hat{g}_T + H_T b_T - h_T) \\ &+ (B_T\hat{g}_T + b_T)' H_T (B_T \hat{g}_T + b_T) \\ &- 2(B_T\hat{g}_T + b_T)' h_T + d_T \end{aligned} \quad (3.13)$$

which is a quadratic function of Y_{T-1}.

Equations (3.10)–(3.13) give the solution to the optimal control problem at time T subject to the constraints (3.1). Now, considering the problem from period $T - 1$ onwards, and by the Bellman principle of optimality, we need only to minimise

$$\begin{aligned} W_{T-1} &= Y'_{T-1} M_{T-1} Y_{T-1} - 2Y'_{T-1} M_{T-1} a_{T-1} \\ &+ a'_{T-1} M_{T-1} a_{T-1} + W_T^* \end{aligned} \quad (3.14)$$

with respect to x_{T-1} subject to the new set of constraints defined at time $T - 1$:

$$F_{T-1}x_{T-1} = f_{T-1} \tag{3.15}$$

Replacing W_T^* in (3.14) by (3.13) will give

$$W_{T-1} = Y'_{T-1}H_{T-1}Y_{T-1} - 2Y'_{T-1}h_{T-1} + d_{T-1} \tag{3.16}$$

where

$$\begin{aligned} H_{T-1} &= M_{T-1} + (A_T + B_T\hat{G}_T)'H_T(A_T + B_T\hat{G}_T) \\ &= M_{T-1} + A'_TH_T(A_T + B_T\hat{G}_T) \end{aligned} \tag{3.17}$$

$$\begin{aligned} h_{T-1} &= M_{T-1}a_{T-1} + (A_T + B_T\hat{G}_T)'(h_T - H_TB_T\hat{g}_T \\ &\quad - H_Tb_T) \\ &= M_{T-1}a_{T-1} + A'_T(h_T - H_TB_T\hat{g}_T - H_Tb_T) \end{aligned} \tag{3.18}$$

$$\begin{aligned} d_{T-1} &= a'_{T-1}M_{T-1}a_{T-1} + (B_T\hat{g}_T + b_T)'H_T(B_T\hat{g}_T + b_T) \\ &\quad - 2(B_T\hat{g}_T + b_T)'h_T + d_T \end{aligned} \tag{3.19}$$

Noticing that the expression (3.16), which is to be minimised with respect to x_{T-1} subject to (3.15), is of the same form as the expression (3.2), we can repeat the process from (3.2) to (3.19) and replace the subscript $T - 1$ for T. This can then be repeated for each time period. Thus, the solution to the T-period control problem is complete and yields the optimal control rule at each period as

$$x_t^* = \hat{G}_tY_{t-1} + \hat{g}_t, \qquad t = 1, \ldots, T \tag{3.20}$$

Here the feedback matrix $\hat{G}_t$ and forcing vector $\hat{g}_t$ are given by

$$\hat{G}_t = G_t - D_t^{-1}F_t'P_t^{-1}F_tG_t \tag{3.21}$$

$$\hat{g}_t = g_t - D_t^{-1}F_t'P_t^{-1}F_tg_t + D_t^{-1}F_t'P_t^{-1}f_t \tag{3.22}$$

where

$$G_t = -D_t^{-1}B_t'H_tA_t \tag{3.23}$$

$$g_t = D_t^{-1}B_t'(h_t - H_tb_t) \tag{3.24}$$

Also

$$\begin{aligned} H_{t-1} &= M_{t-1} + A_t' H_t (A_t + B_t \hat{G}_t) \\ H_T &= M_T \end{aligned} \tag{3.25}$$

$$\begin{aligned} h_{t-1} &= M_{t-1} a_{t-1} + A_t' (h_t - H_t B_t \hat{g}_t - H_t b_t) \\ h_T &= M_T a_T \end{aligned} \tag{3.26}$$

form a pair of vector–matrix Ricatti equations. The quantities $\hat{G}_t$ and $\hat{g}_t$ are independent of the control solution and can be computed recursively beforehand from $t = T$ to $t = 1$. The optimal control x_t^* may be evaluated in the following steps, which are similar to the problem without constraints on the control variables:

(a) With the initial condition $H_T = M_T$, use the equations (3.7), (3.23), (3.21) and (3.25) to yield, consecutively backwards in time, the quantities $D_T, G_T, \hat{G}_T, H_{T-1}, D_{T-1}, \ldots, \hat{G}_1$.
(b) Then, with the initial condition $h_T = M_T a_T$, use the equations (3.24), (3.22) and (3.26) to obtain the quantities $g_T, \hat{g}_T, h_{T-1}, g_{T-1}, \ldots, \hat{g}_1$.
(c) Finally, with $\hat{G}_t$ and $\hat{g}_t$ computed, the optimal control x_t^* is set by the linear feedback control rule (3.20) dependent on Y_{t-1}.

The minimum value of the objective function (2.15) is given by (3.13) with the subscript T replaced by 1. The quantity d_1 is obtained by solving (3.19) backwards in time from the terminal condition $d_T = a_T' M_T a_T$.

This new constrained control solution differs from the unconstrained case in the structure of Equations (3.21), (3.22), (3.25) and (3.26). However, from the computational point of view, the presence of constraints (3.1) does not pose any real difficulty since the optimal control rule (3.20) is of the same recursive form as (2.29) and can be easily implemented for machine calculation. Notice that, in the case that $F_t = 0$ for $t = 1, \ldots, T$, then (3.20) reduces to (2.29), as would be expected.

It has been pointed out in Chapter 2 that, for the unconstrained control problem, the principle of certainty equivalence and the separation theorem hold in the stochastic case as well as the case with incomplete state information. This result also holds when the control variables are subject to linear equality constraints (3.1) (see Tan, 1979a,b). To demonstrate this, note that the presence of the k linear constraints (3.1) at period t implies that all feasible control policies must lie on an $(m - k)$-dimensional hyperplane in the m-dimensional control space. Now consider partitioning (3.1) into

$$F_t^1 x_t^1 + F_t^2 x_t^2 = f_t \tag{3.27}$$

where F_t^1 is $k \times m - k$, F_t^2 is $k \times k$, $\dim(x_t^1) = m - k$ and $\dim(x_t^2) = k$. As the rank of $(F_t) = k$, we can assume, without loss of generality, that F_t^2 is non-singular. For any given x_t^1 the value of x_t^2 must be

$$\begin{aligned} x_t^2 &= (F_t^2)^{-1} f_t - (F_t^2)^{-1} F_t^1 x_t^1 \\ &= q_t + Q_t x_t^1 \end{aligned} \tag{3.28}$$

if (x_t^1, x_t^2) are to satisfy the constraints (3.27). Thus all feasible control policies are given by

$$x_t = \begin{bmatrix} 0 \\ q_t \end{bmatrix} + \begin{bmatrix} I_{m-k} \\ Q_t \end{bmatrix} x_t^1 \tag{3.29}$$

where x_t^1 is arbitrary. From (3.29), the linear dynamic model (2.13) can be written as a function of x_t^1 which is not affected by the constraints. Hence it follows immediately that certainty equivalence and the separation theorem hold for this modified control problem with linear equality constraints on the controllers. This implies that the optimal solutions in cases corresponding to equality-constrained stochastic control problems, and in cases where the solution to the optimal equality-constrained control problem with only indirectly measurable states, can both be readily obtained from the solution to the deterministic case.

3.3 Optimal control with non-negativity constraints: deterministic case

The previous section has shown that the presence of linear constraints on the control variables does not pose any real difficulty in obtaining the optimal control solution. The solution can still be found analytically with appropriate modifications to the Ricatti system for the basic control problem. However, when *inequality* constraints on the controls are added it is no longer true that analytic solutions can be found. Nevertheless, the true optimal solution can still be found in non-analytic form and is given below. Formally, the optimal control problem in this case may be stated as follows:

Given the linear dynamic model (2.13) *with the initial condition* Y_0, *the non-negativity-constrained optimal control problem is to find the* *set of control variables $x_1, \ldots, x_T$ which minimises the quadratic objective function* (2.15) *when the control variables x_t are subject to the constraints*

$$F_t x_t = f_t, \qquad t = 1, \ldots, T \tag{3.30a}$$

$$x_t \geq 0, \qquad t = 1, \ldots, T \tag{3.30b}$$

When the constraints (3.30) are absent, the optimal control solution is given in Chapter 2. When only the constraints (3.30a) are present, the optimal control solution is derived in the previous section. The common feature of both solutions is that they are analytic and permit a recursive calculation of the optimal control vectors. This is no longer true when the constraints (3.30b) are added. In general, as stated by Jacobs (1974), it is not possible to obtain analytic solutions to problems with inequality constraints. Nonetheless, an optimal solution can still be obtained in some cases and two methods are proposed below which require an extra step calling for an iterative adjustment of the solution to achieve feasibility. The rationale of both of these methods is to use analytical solutions as far as possible.

Method 1: dual simplex method In this first method of solving non-negativity-constrained spatial systems the strategy followed is to solve the T-time period problem simultaneously. In effect this reduces the T-period problem to a one-period problem.

The strategy for determining the solution is in two parts: first, the problem is solved with only the linear equality constraints on the controls; secondly, if the resulting solution is not feasible, then the solution is adjusted iteratively using the dual simplex method of linear programming (Hadley 1962). The first part of the solution can be carried out by using the method of the previous section. However, since the second part of the solution requires all controls to be adjusted simultaneously, the problem is reformulated as a static optimisation problem.

The following definitions are used to recast the T-stage problem as a one-stage problem:

$$y = (Y'_1, \ldots, Y'_T, x'_1, \ldots, x'_T)'$$

$$J_1(y) = \begin{bmatrix} A_1 Y_0 + B_1 x_1 + b_1 - Y_1 \\ \vdots \\ A_T Y_{T-1} + B_T x_T + b_T - Y_T \end{bmatrix}$$

$$J_2(y) = \begin{bmatrix} F_1x_1 - f_1 \\ \vdots \\ F_Tx_T - f_T \end{bmatrix}$$

$$J_3(y) = (x'_1, \ldots, x'_T)' = X$$

The static programming problem is that of minimising (2.15) subject to the constraints $J_1(y) = 0$, $J_2(y) = 0$ and $J_3(y) \geq 0$.

STEP 1 Form the Lagrangian

$$L = \tfrac{1}{2}W + \alpha' J_1(y) + \lambda' J_2(y) - \beta' J_3(y),$$

where α, λ and β are vectors of Lagrange multipliers of appropriate dimensions. Both α and λ are unrestricted but β must be non-negative. Note that the objective function and the constraints are convex. Hence, from the Kuhn–Tucker conditions, which are necessary and sufficient (Pindyck 1973), we have after some simplifications

$$\tilde{B}X + AY_0 - h + F'\lambda - \beta = 0 \tag{3.31}$$

where $F = \text{diag}(F_1, \ldots, F_T)$ is a block diagonal matrix of dimension $Tk \times Tm$,

$$A = \begin{bmatrix} B'_1M_1A_1 + B'_1A'_2M_2A_2A_1 + \ldots + B'_1A'_2\ldots A'_TM_TA_T\ldots A_1 \\ \vdots \\ B'_{T-1}M_{T-1}A_{T-1}\ldots A_1 + B'_{T-1}A'_TM_TA_T\ldots A_1 \\ B'_TM_TA_T\ldots A_1 \end{bmatrix}$$

which has dimension $Tm \times (pn + qm)$

$$h = \begin{bmatrix} B'_1M_1 & B'_1A'_2M_2 & \ldots & B'_1A'_2 & \ldots & A'_TM_T \\ & B'_2M_2 & \ldots & B'_2A'_3 & \ldots & A'_TM_T \\ & & \cdot & & & \vdots \\ & \mathbf{0} & & \cdot & & \\ & & & & \cdot & B'_TM_T \end{bmatrix}$$

$$\times \left(\begin{bmatrix} a_1 \\ a_2 \\ \vdots \\ a_T \end{bmatrix} - \begin{bmatrix} b_1 \\ A_2b_1 + b_2 \\ \vdots \\ A_T\ldots A_2b_1 + A_T\ldots A_3b_2 + \ldots + b_T \end{bmatrix} \right)$$

which has dimension $Tm \times 1$. The matrix $\tilde{B}$ is symmetric of dimension $Tm \times Tm$ and its elements are given by

$$\tilde{B} = (b_{ij}), \qquad i, j = 1, \ldots, T,$$

where

$$b_{ii} = B_i' C_i B_i$$

$$b_{ij} = B_i' C_i A_i \ldots A_{j+1} B_j \quad \text{for} \quad i > j$$

The C_i are found recursively from the following equations:

$$C_T = M_T$$

$$C_{i-1} = M_{i-1} + A_i' C_i A_i, \qquad i = 1, \ldots, T \tag{3.32}$$

Now, assuming that $\tilde{B}$ is non-singular, it follows from (3.31), and the equality constraints on X, that the Lagrangian multiplier

$$\lambda = P^{-1}(FG_s Y_0 + Fg_s - f + F\tilde{B}^{-1}\beta) \tag{3.33}$$

where f is a $Tk \times 1$ vector given by $f = (f_1', \ldots, f_T')$,

$$P = F\tilde{B}^{-1}F', \quad G_s = -\tilde{B}^{-1}A \quad \text{and} \quad g_s = \tilde{B}^{-1}h.$$

Substituting (3.33) into (3.31) we obtain

$$X = \hat{G}_s Y_0 + \hat{g}_s + \bar{B}\beta \geqslant 0 \tag{3.34}$$

where we have set

$$\hat{G}_s = G_s - \tilde{B}^{-1}F'P^{-1}FG_s$$

$$\hat{g}_s = g_s - \tilde{B}^{-1}F'P^{-1}Fg_s + \tilde{B}^{-1}F'P^{-1}f$$

$$\bar{B} = \tilde{B}^{-1} - \tilde{B}^{-1}F'P^{-1}F\tilde{B}^{-1}$$

Note that the optimal controls X^* and optimal Lagrange multipliers β^* must satisfy (3.34) together with the following feasibility conditions obtained from Kuhn–Tucker theory:

$$X^* \geqslant 0; \quad \beta^* \geqslant 0 \quad \text{and} \quad \beta^{*\prime} X^* = 0 \tag{3.35}$$

Observe that when the constraints (3.30b) are absent then (3.34) reduces to

$$X^\circ = (x_1^{\circ\prime}, \ldots, x_T^{\circ\prime})' = \hat{G}_s Y_0 + \hat{g}_s = \bar{G}_s \tag{3.36}$$

which is exactly the same as the optimal control solution obtained for the equality constrained case.

STEP 2 When (3.36) does not satisfy (3.35), the second part of the strategy considers the following programming problem:

$$\text{maximise } z = 1_v'(FX - f) \tag{3.37}$$

subject to

$$\begin{aligned} X - \bar{B}\beta &= \bar{G}_s \\ X &\geqslant 0 \\ \beta &\geqslant 0 \\ \beta' X &= 0 \end{aligned}$$

where 1_v is a $Tk \times 1$ vector whose elements are all unity. Except for the restriction $\beta = 0$ the problem is linear. An initial basic solution to (3.37) is $X_b = \bar{G}_s$. Of course, we assume that $\bar{G}_s \ngeq 0$ otherwise the control problem with non-negativity constraints is solved. Thus the initial basic solution to the primal problem is infeasible. However, it may be checked that it is dual feasible, so that the dual simplex method can be used to solve this problem (3.37). However, because of the constraint $\beta' X = 0$, it is necessary to introduce a variation to this algorithm. The variation is similar to that given in Wolfe's (1959) algorithm for solving quadratic programming problems, namely that if $X_i > 0$ then β_i is not allowed to enter the basis and vice versa. Initially $z = 0$ and remains at this level at each iteration of the dual simplex method. This ensures that the set of linear equality constraints (3.30a) is satisfied and also maintains dual feasibility of the iterations. For well defined problems, the algorithm will terminate to an optimal solution X^* and β^* in a finite number of steps.

If the matrix $\bar{B}$ is of full rank (i.e. Tm) then, from (3.34), $\beta^* \geqslant -\bar{B}^{-1}\bar{G}_s$ and, together with the feasibility conditions (3.35), the problem of finding β^* and X^* could be solved directly. When there are no equality constraints on controls (that is, $k = 0$), then $\bar{B}$ is non-singular, otherwise it may be shown (Rao 1973) that $\bar{B}$ is of rank $T(m - k)$. Hence, β^* and X^* are determined using the dual simplex method.

The procedure to obtain the truly optimal solution may, then, be summarised as follows:

Step 1 Solve the optimal control problem with only the linear equality constraints on the control variables (3.30a) present in exactly the same way as given in Section 3.2. Denote this first part solution by $x_1^{\circ}, \ldots, x_T^{\circ}$.

Step 2 If $x_1^{\circ}, \ldots, x_T^{\circ}$ do not satisfy the non-negativity constraints on the control variables (3.30b), they are adjusted simultaneously for feasibility by solving the programming problem (3.37) using the dual simplex algorithm of linear programming.

The above strategy gives a truly optimal solution to systems that are deterministic. Extensions to the stochastic case are discussed in the next section, and to the spatial case in Chapter 6.

AN EXAMPLE To see how the optimal solutions for the various problems are actually calculated, a simple example is given below of a two-region economy, which illustrates most of the relevant features of the various solutions. This two-region example is assumed to have the following deterministic linear dynamic equation:

$$Y_t = \begin{bmatrix} 2 & 1 \\ 0 & 2 \end{bmatrix} Y_{t-1} + \begin{bmatrix} 1 & 0 \\ 2 & 1 \end{bmatrix} x_t$$

where Y_t and x_t are 2×1 vectors of state and control variables, respectively. Assume that we have a two-stage problem, i.e. $T = 2$ and with initial condition $Y_0 = (3, -1)'$, targets set as $a_1 = (5, 3)'$, $a_2 = (8, 6)'$ and the quadratic cost weights in Equation (2.15) are given by $M_1 = M_2 = I$. For the constraints (3.30) we set $F_1 = F_2 = (1, 1)$ and $f_1 = 5$, $f_2 = 8$. If Y_t represents two important socio-economic indicators and x_t denotes a control vector such as the amount of cash grant allocated to each of the two regions, then the constraints f_i specified above stipulate that, in the first period, the total control setting (the total cash available for distribution) must not exceed five units, while in the second period the constraint is that the total control setting should not exceed eight units. Also, for both periods money cannot be taken from either region. This interpretation of the example highlights the basic idea of allocating cash grants between two levels of government in a national economy, which is discussed in more detail in Chapter 7.

In considering the above numerical example, it is useful to compare the optimal solutions to the problem without any constraints (unconstrained problem) with the control problem, modified either by equality

Table 3.2 Optimal solutions to unconstrained, equality-constrained and non-negativity-constrained control problems.

	Optimal x_t^* at $t = 1$	Optimal x_t^* at $t = 2$
basic problem (unconstrained)	$\begin{bmatrix} 0 \\ 5 \end{bmatrix}$	$\begin{bmatrix} -5 \\ 10 \end{bmatrix}$
modified control (equality constrained)	$\begin{bmatrix} 0.6 \\ 4.4 \end{bmatrix}$	$\begin{bmatrix} -8 \\ 16 \end{bmatrix}$
modified control (equality and non-negativity constrained)	$\begin{bmatrix} 0 \\ 5 \end{bmatrix}$	$\begin{bmatrix} 0 \\ 8 \end{bmatrix}$

constraints or by both equality and non-negativity constraints. The optimal solutions to each of the three problems are summarised in Table 3.2.

Clearly, the introduction of constraints gives significantly different results. The calculations for these solutions are given as follows. For the unconstrained problem,

$$G_2 = \begin{bmatrix} -2 & -1 \\ 4 & 0 \end{bmatrix}, \quad G_1 = \begin{bmatrix} -2 & -1 \\ 4 & 0 \end{bmatrix}$$

$$g_2 = \begin{bmatrix} 8 \\ -10 \end{bmatrix}, \quad g_1 = \begin{bmatrix} 5 \\ -7 \end{bmatrix}$$

Using these results, together with the initial condition Y_0, the optimal controls can be readily derived. For the equality-constrained problem, performing the calculations according to the equations given in Section 3.2, we get

$$\hat{G}_2 = \begin{bmatrix} -1 & -1.5 \\ 1 & 1.5 \end{bmatrix}, \quad \hat{G}_1 = \begin{bmatrix} -1.6 & -1.2 \\ 1.6 & 1.2 \end{bmatrix}$$

$$\hat{g}_2 = \begin{bmatrix} 3 \\ 5 \end{bmatrix}, \quad \hat{g}_1 = \begin{bmatrix} 4.2 \\ 0.8 \end{bmatrix}$$

The optimal control vector for each time period is then determined from Equation (3.20).

To find the optimal control solution for the non-negativity-constrained

Tableau 3.1 Initial tableau.

		β_{11}	β_{12}	β_{21}	β_{22}
	0	0	0	0	0
x_{11}	3/5	−2/5	2/5	1	−1
x_{12}	22/5	2/5	−2/5	−1	1
x_{21}	−8	1	−1	(−3)	3
x_{22}	16	−1	1	3	−3
β_{11}	0	−1	0	0	0
β_{12}	0	0	−1	0	0
β_{21}	0	0	0	−1	0
β_{22}	0	0	0	0	−1

problem, it is necessary to evaluate the matrix $\bar{B}$. This is given by

$$\bar{B} = \frac{1}{5}\begin{bmatrix} 2 & -2 & -5 & 5 \\ -2 & 2 & 5 & -5 \\ -5 & 5 & 15 & -15 \\ 5 & -5 & -15 & 15 \end{bmatrix}$$

and the initial tableau for the programming problem (3.37) is given by Tableau 3.1. The pivot element for the dual simplex algorithm is shown circled. The basic variable x_{21} now leaves the basis and its place is taken by the non-basic variable β_{21}. The tableau after this iteration is shown by Tableau 3.2. Again the pivot element is shown circled. Now, x_{11} leaves the basis and is replaced by β_{11}. The second iteration yields Tableau 3.3. As the first column of Tableau 3.3 does not contain any negative elements, this is the final tableau. Hence, we have found the

Tableau 3.2 Tableau after first iteration.

		β_{11}	β_{12}	x_{21}	β_{22}
	0	0	0	0	0
x_{11}	−31/15	(−1/15)	1/15	1/3	0
x_{12}	106/15	1/15	−1/15	−1/3	0
x_{21}	0	0	0	−1	0
x_{22}	8	0	0	1	0
β_{11}	0	−1	0	0	0
β_{12}	0	0	−1	0	0
β_{21}	8/3	−1/3	1/3	−1/3	−1
β_{22}	0	0	0	0	−1

Tableau 3.3 Final tableau after two iterations.

		x_{11}	β_{12}	x_{21}	β_{22}
	0	0	0	0	0
x_{11}	0				
x_{12}	5				
x_{21}	0				
x_{22}	8				
β_{11}	31				
β_{12}	0				
β_{21}	13				
β_{22}	0				

optimal solution to be

$$X^* = \begin{bmatrix} 0 \\ 5 \\ 0 \\ 8 \end{bmatrix} \quad \text{and} \quad \beta^* = \begin{bmatrix} 31 \\ 0 \\ 13 \\ 0 \end{bmatrix}$$

It is not necessary to determine the remaining entries of Tableau 3.3 because this is the final tableau yielding the optimal solution satisfying the constraints.

Method 2: Imposing constraints The use of the dual simplex algorithm for solution of the inequality-constrained problem is time-consuming and cumbersome. As a result, an alternative solution developed by Tan (1979a,b) is by a method of imposing constraints. This is also a two-step procedure which can be summarised as follows:

STEP 1 With only the equality constraint (3.30a) present, the optimal control solution is given, as discussed above (Section 3.2) by

$$x_t^\circ = \hat{G}_t y_{t-1} + \hat{g}_t \tag{3.38}$$

where $\hat{G}_t$ and $\hat{g}_t$ are defined by (3.21) and (3.22).

If this solution path does satisfy the non-negativity constraints (3.30b), then it is also the optimal solution to the control problem stated above. However, if this solution path does not satisfy the constraints (3.30b), then the second step of the solution must be followed.

STEP 2 This step is stimulated by the property of the dual simplex algorithm noted earlier, that the basic variable with the largest negative value is made non-basic (i.e. it is set to the value of zero). When this basic variable is a control variable then making it non-basic amounts to adding another control constraint namely that this control variable is equal to zero.

Using this property the method of imposing constraints consists of a three-step strategy:

(a) Set largest negative control variable to zero (its non-basic value).
(b) Re-solve the T-stage control problem as an equality-constrained problem.
(c) Check whether all control variables are non-negative. If not, set largest negative control variable to zero as in step (a) and repeat until all control variables are non-negative, and hence that the constraint (3.30b) is satisfied.

It is not necessarily true that this control solution is the true optimum. This is because once a control variable is set to zero, it remains so throughout the analysis. However, the solution obtained by this method of imposing constraints does take into account the multistage nature of the problem and should offer a close approximation to the true optimum in most cases; and it does yield the identical optimal solution to that obtained from the dual simplex method of Equation (3.38).

3.4 Optimal control with non-negativity constraints: stochastic case

The solution strategy developed above cannot be applied directly to the case in which we have a stochastic control problem, i.e. when there is an additive vector of random disturbances present, as in (2.30), and there are inequality constraints. The reason for this is that in the stochastic situation it is not possible to determine numerically in period 1 the controls $x_1^\circ, \ldots, x_T^\circ$ which are the solutions to the problem with only linear equality constraints since these are not known with certainty. Knowledge of these controls $x_1^\circ, \ldots, x_T^\circ$ is needed for the adjustment mechanism. Despite this problem the control strategy proposed in Section 3.3 can be adapted to yield feasible solutions which are near optimal. This may be achieved as follows.

When the additive random term u_t is set to its expected value (i.e. to zero) for all t, the solution method given above allows the expected trajectory of the controls, with only equality constraints (3.30a) present, to be determined in the first period. Denote these by $x_1^e, \ldots, x_T^e$. The solution is simultaneously adjusted so as to satisfy the non-negativity constraints (3.30b). The adjustment is carried out by the dual simplex method described above. As the model parameters are all known with certainty, the matrix $\bar{B}$ of (3.37) need not be altered for this adjustment step. Denote the adjusted feasible solutions by $x_1^a, \ldots, x_T^a$. These represent the expected trajectory of the controls subject to linear as well as non-negativity constraints. The idea of setting the random term u_t to its expected value in order to get the solution path of x_t is an obvious one and has been used by Chow and Megdal (1978) for the control of a large-scale econometric model without any constraint on the controllers.

In practice, only x_1^a is applied and when Y_1 becomes known it is taken as the initial condition for getting the solution to the $(T - 1)$-period problem. In other words, at time $t = 2$, setting the random terms $u_2, \ldots, u_T$ to zero, determine the controls $x_2^e, \ldots, x_T^e$ when only the linear constraints (3.30a) are imposed. This solution is then adjusted to satisfy the non-negativity constraints, yielding $x_2^a, \ldots, x_T^a$. Again at time $t = 2$, only x_2^a is applied and the process is repeated until x_T^a is found. Even though at each period t we calculate the near-optimal feasible solutions $x_t^a, \ldots, x_T^a$ for the $(T - t + 1)$-stage control problem, only the 'first-period' control x_t^a is actually used. The control x_t^a does take into consideration the multistage nature of the problem and uses all information available at time t. This procedure is summarised by the following steps:

(a) At time t, solve the $(T - t + 1)$-stage stochastic optimal control problem with only the linear equality control constraints (3.30a) present by the method described in Section 3.2. Setting the additive random error term present in the linear dynamic model to its expected value (i.e. zero), determine the control setting $x_t^e, \ldots, x_T^e$.

(b) If $x_t^e, \ldots, x_T^e$ do not satisfy the non-negativity control constraints (3.30b), then $x_t^e, \ldots, x_T^e$ need to be adjusted for feasibility. This can be done by solving the programming problem (3.37), as discussed in Section 3.3. Denote the resulting adjusted solution by $x_t^a, \ldots, x_T^a$.

It should be noted that the adjustment to satisfy the non-negativity constraints (3.30b) may be carried out by the method of imposing constraints described previously.

3.5 *Conclusion*

This chapter has been concerned with developing algorithms to solve control problems with linear equality and non-negativity constraints on the control variables. In Section 3.2 we have considered the situation where only linear equality control constraints are present. In this case it turns out that the optimal control settings can still be given in a linear feedback form. Furthermore, it is shown that the separation theorem and the principle of certainty equivalence are both also valid. However, these results do not carry over to the case where non-negativity control constraints are added. For a deterministic linear dynamic model, Section 3.3 gives a method by which to find the true optimal solution to the control problem with non-negativity control constraints. The method is made up of two parts. The first part solves the problem with only linear equality constraints present, as in the discussion of Section 3.2. If this first part solution does not satisfy the non-negativity control constraints, then a second part of the solution must be utilised to adjust the first part solution for feasibility. It turns out that the dual simplex method provides the necessary adjustment machinery. An alternative method of imposing constraints has also been described and this allows the adjustment of the control settings to satisfy non-negativity. Although attempts have been made to solve the control problem considered in this chapter (e.g. Ritch 1973, Evtushenko & MacKinnon 1975, Mehra 1975), these previous methods endeavour to find a local optimum. There is no guarantee that these methods will yield the true global optimal solution. Furthermore, most previous solutions have been developed only for deterministic linear dynamic models. In contrast, the two-part solution strategy given here can be adapted to the stochastic situation, where the linear dynamic model contains an additive random error term, as discussed in Section 3.3.

Throughout this chapter it has been assumed that the model parameters are known with certainty. The case of unknown model parameters is a more typical practical occurrence and provides extra difficulties in obtaining optimal solutions. Solution strategies for such uncertain parameter situations are discussed in Chapters 4 and 5.

4 *Spatial control with constant but unknown parameters*

4.1 Joint estimation and control

One of the most frequent practical situations is the desire to control when many of the constants and coefficients in the system are unknown. It is then said that control is subject to uncertainty.

Three types of uncertainty may be present. First, an additive random error term may corrupt the state equation. Secondly, the measurement equation may include an additive random error term. Thirdly, the parameters of the linear model may be unknown. So far, the optimal control problems considered in this book have included consideration of the first two sources of uncertainty but have ignored the third. The purpose of this chapter is to derive control policies for spatial problems in the presence of the third source of uncertainty, unknown parameters, for both unconstrained and constrained problems. The objective function used is the familiar quadratic.

There are two possible interpretations for uncertainty in system parameters: first, that the true linear dynamic model has constant but unknown parameters; secondly, that the unknown parameters are time-varying. In general, the solutions obtained when the parameters are uncertain are only approximately optimal (Chow 1976b) and simplifying assumptions have to be made to avoid the resulting solution becoming excessively difficult to implement. Generally, it is preferable and more practical to employ sub-optimal control rules, that are readily evaluated, than truly optimal rules that either require a great deal of effort to compute, or in some cases cannot be computed at all. For control problems with stochastic parameters, no truly optimal solution has yet been obtained. As a result the emphasis is normally placed on obtaining solutions that are near optimal but lend themselves to easy evaluation. In this chapter we shall consider constant but unknown parameters. The case of unknown time-varying parameters is the subject of the next chapter.

The need to consider such uncertainties in model parameters when applying optimal control theory to the social sciences is self-explanatory: it is unusual for realistic models in both environmental and social sciences to come with known parameters. The normal situation is for time-series or spatial cross sections of data to be available with only a little prior knowledge of the form of any model and its parameters. Hence, there is in many practical control problems a need to estimate model parameters, and even in some cases to specify the nature of the model structure.

In macro-economic and environmental applications the traditional approach to handling uncertainty in the model parameters is, first, to estimate these parameters and, then, to reformulate the model using these estimates in place of the unknown parameters and treating the reformulated model as completely deterministic; see, for example, Pindyck (1973). Attempts have been made to treat the problem with unknown parameters more directly (e.g. Brainard 1967, Kendrick & Majors 1974, Chow 1975, Shupp 1976, Turnovsky 1977a,b). These interpretations of the problem have varied. For example, Brainard's analysis is static while Shupp extended results to a dynamic situation in the presence of instrument costs, i.e. the manipulation of the control variables involve costs and these appear explicitly in the objective function. The analysis by Kendrick and Majors only deals with noise in the coefficient of the state variables. More general treatments are given by Chow (1975) and Turnovsky (1977a,b). However, constraints on the controllers are omitted in these studies and the purpose of this chapter is to develop feasible solutions to problems where such constraints are present.

It is assumed below that the system has not been subject to control *before* the derivation of the policy control solutions which are to be derived. There are special difficulties in the respecification of control solutions in systems which have been previously subject to control and in which model parameters are unknown. These are discussed, for example, in Bennett (1979, pp. 600–3).

We are concerned here solely with control in systems in which there has been no prior control, but in which system parameters and/or structure are unknown. This leads us to two situations: control with unknown system structure and control with unknown system parameters.

Control with unknown system structure When system structure is unknown, or only partially known, there are two main situations which occur: first, when there are restrictions on system observability; secondly, when there are restrictions on the system controllability. When the system variables are subject to theoretical observability and controllability restrictions, the control problem can be solved by reconstructing the system state variables. For deterministic systems this can be carried out by using the *observer* method (see Luenberger 1964). For stochastic systems the Kalman filter (Kalman 1960, Bryson & Ho

1969) can be used. Most of the discussion below will assume that the system under study is observable and controllable. However, the Kalman filter also provides a means for solving other control problems and this is discussed extensively below.

Control with unknown system parameters In this situation the structure of the system model is assumed to be known, but the coefficients of the model are unknown. As a consequence, an estimation problem results. A wide variety of methods can be applied to the solution of this problem but, because of its attractive properties and wide application, attention is restricted here to least-squares (OLS) estimators. These have the advantage that under relatively weak conditions, they are 'best' in the sense that they are minimum variance linear unbiased estimators. Furthermore, they permit easy computation and are readily modified to yield recursive estimators, of the Kalman filter form, when new information is available (Plackett 1950, Phillips 1977).

The minimum variance unbiased estimation property derives from the important result that minimum mean square error (MMS error) estimates or forecasts are a linear combination of estimated inputs and outputs, and therefore any linear combination of the MMS error forecasts will also be MMS error. As a result MMS error forecasts will also yield MMS error policy and control solutions. With reference to OLS estimation of control strategies, we also have the important result (Åstrom & Wittenmark 1972) that, in general, even biased estimates derived from OLS give optimal control strategies using feedback from forecasts. Hence we have the result that MMS forecasts and parameter estimates yield optimal controllers. This enables the use of least squares as a parameter estimation device, as a forecasting method, and as an optimal controller. There are two caveats, however. First, identification and estimation are subject to difficulties and limitations if there is already some form of feedback control device affecting the input–output record. Secondly, the use of OLS or other estimators does *not* imply that a separation theorem always exists; that is, parameter estimates cannot necessarily be substituted into control solutions such that the estimation and control problems are independent of one another (Åstrom & Wittenmark 1971).

Separability Considerable discussion has surrounded the issue of whether a separation theorem does exist which, under certain conditions, allows control of systems independent of the estimation of the system parameters. There are three possible approaches to control in situations in which the parameters must be estimated, as shown in Figure 4.1. In the first case the separation principle is assumed to hold and there is separate identification and control. In the second case there

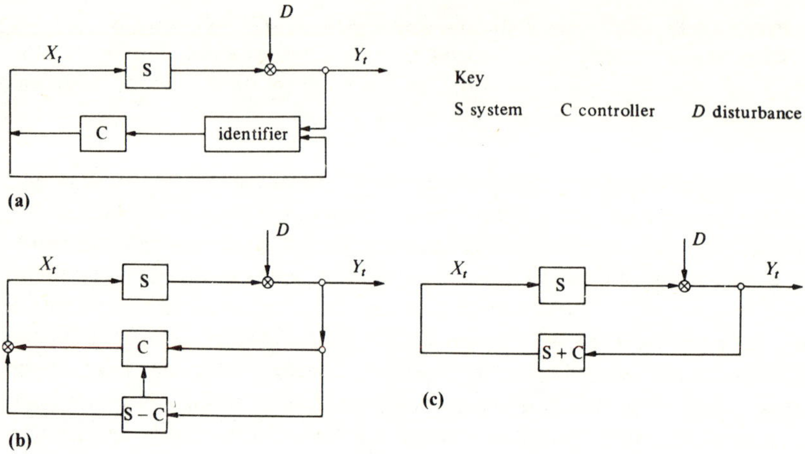

Figure 4.1 Three categories of approach to joint estimation and control: (a) the separation principle; (b) estimation of combined system plus controller; (c) estimation of controls only (after Strejc 1973, Figs 10.1–3, and Bennett 1979, Fig. 8.20).

is estimation of a combined system-plus-control model. This is usually adaptive so that direct estimation of control parameters (with use of input–output data) is used to make changes to the system controller. The third possibility is to ignore the parameter structure of the system altogether and to estimate only a control model by using just the output data. No direct identification or estimation is then required. In control of socio-economic and environmental systems we usually wish to know the 'free' system properties as well as the control system, and hence systems of either the first or second type are considered below.

Where the standard errors of the parameters are very low it may be practicable, in many instances, to assume that the separation theorem does hold. As a rule of thumb, coefficient estimates, which are all significant statistically at significance levels appropriate to the problem, *plus* the property that the error variance is less than 5–10% of the system total variance, may usually be taken as appropriate (see, for example, Eykhoff 1974, and the discussion contributions in Eykhoff 1973).

When these properties do not hold, as is common in many practical social, economic and environmental situations, then the second route shown in Figure 4.1 can be followed: that of a combined estimator and controller. There are two main forms of such controllers: first, the Åstrom and Wittenmark (1971, 1973) *self-tuning regulator*, which yields asymptotic optimal control and parameter estimates but only in univariate systems; second, Chow's (1975) *iterative scheme*, which yields sub-optimal results but can be applied to multivariate systems. Other discussions of this problem and possible solutions are given by Prescott

(1972), Brito (1973), Tse (1974), MacRae (1975) and Rausser and Freebairn (1978).

The Åstrom and Wittenmark self-tuning regulator Åstrom and Wittenmark recognise the property that continued system control and system operation both generate further information as to the structure of the system. By using this information it is possible to implement joint parameter estimation, identification and control; with the use of least-squares parameter estimates, for example. Wieslander and Wittenmark (1971), Åstrom and Wittenmark (1973) and Peterka and Åstrom (1973) give the following self-tuning regulator:

Step 1 Determine the parameter of the system at each time period by OLS, based on all available data.
Step 2 Determine the control rule for the model estimated at step 1.
Step 3 Repeat steps 1 and 2 at each succeeding sampling interval.

The equation for this self-tuning regulator is a form of recursive least-squares algorithm which is a special form of Kalman filter algorithm. This achieves two important optimality properties, given as two theorems by Åstrom and Wittenmark (1972):

THEOREM 1 (separability). *Provided that the OLS identification algorithm emerges, in that the closed-loop system is stationary in second moments (variance and covariance), then the closed-loop system has the properties*

$$E(Y_t, Y_{t+\tau}) = 0, \qquad \tau = k + 1, \ldots, k + p \tag{4.1}$$

$$E(Y_{t+\tau}, X_t) = 0, \qquad \tau = k + 1, \ldots, k + q \tag{4.2}$$

It is sufficient for stationarity (ergodicity) of the closed-loop system that the system is of finite order (p and q) and that the closed-loop system is stable. Thus the control of the present and of the future output will be independent. A separation principle holds only for the recursive formulation.

THEOREM 2 (optimality). *With the assumption that the estimation control algorithm converges, the resultant closed-loop system corresponds to a minimum variance control strategy.*

These separation and optimality principles will hold whatever the form of the close-loop system, provided that it is stable. This result is most important, since if estimates of the system parameters by least-squares regression are available, these can be used to implement optimal controllers even though a separation principle does not exist. Moreover, the updating of the parameter estimates allows the accuracy of the control to be improved at each step. Hence, provided that the OLS estimates of the system converge to the true parameters, the sub-optimal control rule is asymptotically optimal.

The Chow iterative scheme Chow's (1975, Ch. 11) approach to joint parameter estimation and control is derived by recognising from the outset that some modification based on future knowledge of the system is required to take account of parameter uncertainty. This requires minimisation of a modified form of control objective function which, for quadratic costs, Chow suggests can be given by

$$W = \frac{1}{2}\sum_{t=1}^{T} Y_t' M_{tt} Y_t + \sum_{t=1}^{T}\sum_{s<t} Y_t' M_{ts} Y_s + \sum_{t=1}^{T} Y_t' m + k \tag{4.3}$$

The cost matrix M of Equation (2.15) has now been expanded into three terms in Equation (4.3). The first term M_{tt} represents the weights at any time t. The new term M_{ts} represents the weights placed on correlation between past system outputs. The third term m is a vector representing generalised costs, and the term k is an additive constant. It is assumed that each of these matrices and vectors is known or prespecified.

Chow's scheme is complex and results in a non-linear optimisation problem. This is approximated by a series of iterative steps, each corresponding to a chosen tentative path of the future state variables. Alternative approaches are to simplify Chow's method; for example, the lagged cost covariance matrices M_{ts} can be omitted or the method can be applied for a shorter time horizon than T. Whichever method is chosen, however, the results are computationally costly unless the system structure is very simple.

4.2 *Estimation in controlled systems*

The preceding discussion has raised the difficulties and assumptions required in joint estimation and control. In the rest of this book, two alternative approaches to control in systems with parameter uncertainties are presented:

(a) *Separation assumed.* The unknown system parameters are first estimated (e.g. by OLS), then the control solution is determined after assuming these estimates to be known with certainty. This is most apposite when there is low system error variance, low parameter standard errors, and fairly simple models which are constant or nearly constant over the planning period.

(b) *Recursive self-tuning regulator*. A special form of the Åstrom and Wittenmark self-tuning regulator is employed in which parameter estimates are updated during the control period. This method is most apposite in situations in which there is considerable parameter uncertainty, models are complex, or the system parameters are changing rapidly over the planning period. These results are only asymptotically optimal but do yield a good practical approach.

Both approaches to estimation and control can be founded on the same general formulations. In the discussion which follows, the concept of separation is assumed to apply, but in Chapter 5 the results are extended to a form of self-tuning controller, based on the Kalman filter, which is similar to the Åstrom and Wittenmark scheme.

For the case in which separation can be assumed to apply, when the original system, leading to (2.30), is not of first order, and when lagged control variables are present, then the matrices A and B and vector b are given by (2.14). In this case it seems reasonable simply to determine the estimates of the parameter 'elements' of the 'first row' of A, B and b; i.e. to calculate the estimates of the unknown parameters only in the original reduced-form equations.

Adopting this approach, assume that the original reduced-form equations are given by

$$y_t = A_1 y_{t-1} + \ldots + A_p y_{t-p} + B_1 x_{t-1} + \ldots + B_q x_{t-q} + B_0 x_t + \bar{b} w_t + e_t \quad (4.4)$$

where y_t is a vector of dimension $n \times 1$, the control vector x_t is of dimension $m \times 1$ and w_t is a $n \times 1$ vector of exogenous variables not subject to control, which is assumed to be known. Referring back to (2.12), $\bar{b}_t = \bar{b} w_t$ and $\bar{b}$ is the unknown parameter matrix. This treatment includes as a particular case the regression equation with an intercept term. If we consider n as the number of regions, then y_t represents one 'dependent' variable for each of the regions. The matrices A, B and $\bar{b}$ in (4.4) are the unknown parameters that have to be estimated.

Denote the ith row of A_r, B_s and $\bar{b}$ by a_r^i, b_s^i and $\bar{b}^i$, respectively. Then, for the ith region we have, from (4.4), the univariate regression equation

$$y_{it} = a_1^i y_{t-1} + \ldots + a_p^i y_{t-p} + b_1^i x_{t-1} + \ldots + b_q^i x_{t-q} + b_0^i x_t + \bar{b}^i w_t + e_{it}. \quad (4.5)$$

Supposing that there are N observations available to estimate all of the unknown parameters in (4.5), then the linear model may be represented by

$$y^i = X_i \beta_i + e^i \tag{4.6}$$

where $y^i = (y_{i1}, y_{i2}, \ldots, y_{iN})'$ is a $N \times 1$ vector. X_i is the regression matrix which is given by

$$X_i = \left[\begin{array}{ccc|c|ccc|} y_{1,0} & \cdots & y_{n,0} & & y_{1,1-p} & \cdots & y_{n,1-p} \\ \vdots & & \vdots & \cdots & \vdots & & \vdots \\ y_{1,N-1} & \cdots & y_{n,N-1} & & y_{1,N-p} & \cdots & y_{n,N-p} \end{array}\right.$$

$$\begin{array}{ccc|c|ccc|} x_{1,0} & \cdots & x_{m,0} & & x_{1,1-q} & \cdots & x_{m,1-q} \\ \vdots & & \vdots & \cdots & \vdots & & \vdots \\ x_{1,N-1} & \cdots & x_{m,N-1} & & x_{1,N-q} & \cdots & x_{m,N-q} \end{array}$$

$$\left.\begin{array}{ccc|ccc} x_{1,1} & \cdots & x_{m,1} & w_{1,1} & \cdots & w_{n,1} \\ \vdots & & \vdots & \vdots & & \vdots \\ x_{1,N} & \cdots & x_{m,N} & w_{1,N} & \cdots & w_{n,N} \end{array}\right]$$

This is of dimension $N \times [n(p+1) + m(q+1)]$. β_i is the unknown parameter vector of dimension $[n(p+1) + m(q+1)] \times 1$:

$$\beta_i' = (a_1^i, \ldots, a_p^i, b_1^i, \ldots, b_q^i, b_0^i, \bar{b}^i)$$

e^i is the random error vector. Without any loss of generality, the regressor matrix in the ith equation, X_i, may be taken to be of dimension $N \times k_i$ with rank k_i and β_i a $k_i \times 1$ vector of regression coefficients. This makes allowance for situations in which the design matrix (or regressor matrix) may differ between the various regression equations for each of the dependent variables. The system of equations describing the estimation of the unknown parameters, of which (4.6) is one equation, may be written compactly as

$$\begin{pmatrix} y^1 \\ \vdots \\ y^n \end{pmatrix} = \begin{bmatrix} X_1 & 0 & \cdots & 0 \\ \vdots & & & \\ 0 & 0 & \cdots & X_n \end{bmatrix} \begin{pmatrix} \beta_1 \\ \vdots \\ \beta_n \end{pmatrix} + \begin{pmatrix} e^1 \\ \vdots \\ e^n \end{pmatrix}$$

or

$$\underset{(Nn\times 1)}{y} = \underset{(Nn\times k)}{X}\ \underset{(k\times 1)}{\beta} + \underset{(Nn\times 1)}{e} \tag{4.7}$$

where

$$k = \sum_{i=1}^{n} k_i$$

It follows immediately that the model (4.7) may be regarded as seemingly univariate, i.e. as a single-equation regression model.

Before deriving the least-squares estimators, let us assume that the random error terms e^i are each distributed with mean zero and variance σ_{ii} and are uncorrelated over time, i.e.

$$\begin{aligned} E(e^i) &= 0 \\ E(e^i e^{i\prime}) &= \sigma_{ii} I_N \end{aligned} \tag{4.8}$$

Furthermore, we assume that e^i and e^j are not independently distributed. This implies that there is spatial autocorrelation. The covariance matrix of e^i and e^j is

$$E(e^i e^{j\prime}) = E\begin{bmatrix} e_{i1}e_{j1} & e_{i1}e_{j2} & \dots & e_{i1}e_{jN} \\ \vdots & & & \vdots \\ e_{iN}e_{j1} & e_{iN}e_{j2} & \dots & e_{iN}e_{jN} \end{bmatrix}$$

and if there is no temporal autocorrelation and no heteroscedasticity, this may be written as

$$E(e^i e^{j\prime}) = \sigma_{ij} I_N, \qquad i, j = 1, 2, \dots, n \tag{4.9}$$

From (4.8) and (4.9), the covariance matrix of e in (4.7) is then given by

$$\Sigma_e = E(ee') = \begin{bmatrix} \sigma_{11}I_N & \sigma_{12}I_N & \dots & \sigma_{1n}I_N \\ \vdots & \vdots & & \vdots \\ \sigma_{n1}I_N & \sigma_{n2}I_N & \dots & \sigma_{nn}I_N \end{bmatrix} = \Sigma \otimes I_N \tag{4.10}$$

where

$$\Sigma = \begin{bmatrix} \sigma_{11} & \dots & \sigma_{1n} \\ \vdots & & \vdots \\ \sigma_{n1} & \dots & \sigma_{nn} \end{bmatrix}$$

is the $n \times n$ covariance matrix of $(e_{1t}, e_{2t}, \ldots, e_{nt})'$ for any t, and $\otimes$ is the Kronecker product for matrices (Rao 1973). The expression (4.10) implies constant variances and covariances from period to period. If panel data is available, i.e. there is a time series of cross-sectional data, the problem of estimating β has been solved by Zellner (1962) using a method based on Aitken's estimator (generalised least-squares). Specifically, the Aitken estimator is

$$\hat{\beta} = (X'[\Sigma^{-1} \otimes I_N]X)^{-1}X'[\Sigma^{-1} \otimes I_N]y \tag{4.11}$$

However, when the matrix Σ is unknown it is impossible to compute the estimate (4.11). A way round this problem is to replace Σ by any consistent estimator. The usual choice is to use the single-equation least-squares residuals for the appropriate equations. Thus, the matrix $\Sigma = (\sigma_{ij})$ is estimated by

$$s_{ij} = \frac{(y^i - X_i\hat{\beta}_i)'(y^j - X_j\hat{\beta}_j')}{(N - k_i)^{1/2}(N - k_j)^{1/2}}$$

where $\hat{\beta}_i = (X_i'X_i)^{-1}X_i'y^i$ is the usual single-equation ordinary least-squares estimator. Once $\hat{\Sigma} = (s_{ij})$ is obtained, its inverse can be calculated, which is then used to form the estimator

$$\tilde{\beta} = (X'[\hat{\Sigma}^{-1} \otimes I_N]X)^{-1}X'[\hat{\Sigma}^{-1} \otimes I_N]y \tag{4.12}$$

The statistical properties of the Aitken estimator $\hat{\beta}$ given by (4.11) have been investigated for the situation where the design matrix consists of independent non-stochastic variables by Zellner (1962) and by Zellner and Huang (1962), who demonstrate that this 'systems' estimator is at least asymptotically more efficient than the least-squares estimators obtained equation by equation. Furthermore, $\hat{\beta}$ is a best linear unbiased estimate, and it is also a maximum likelihood estimate if it is assumed that the random error terms are normally distributed. The properties of the practical estimator $\tilde{\beta}$ given according to (4.12) have also been derived with reference to the theoretical estimator $\hat{\beta}$. Some of the properties are as follows (Zellner 1962, Zellner & Huang 1962):

(a) $\tilde{\beta} = \hat{\beta} + O(N^{-1})$, where $O(N^{-1})$ denotes a quantity which is of order N^{-1} in probability.
(b) The bias of $\tilde{\beta}$ is at most of order N^{-1}.
(c) $(\tilde{\beta} - \beta)$ and $(\hat{\beta} - \beta)$ have the same asymptotic covariance matrix.

(d) $N^{1/2}(\tilde{\beta} - \beta)$ and $N^{1/2}(\hat{\beta} - \beta)$ have the same asymptotic normal distribution.

These properties (a)–(d) confirm that the practical estimator $\tilde{\beta}$ performs well with respect to the theoretical estimator $\hat{\beta}$.

When the design matrix contains lagged dependent and exogenous variables, it is questionable whether the explanatory variables can be regarded as independent. For such a case, it is still not known whether the properties of the estimators $\tilde{\beta}$ and $\hat{\beta}$ given above still hold. However, in the univariate case, least squares is usually the best estimating technique (Copas 1966, Johnston 1972). Hence, it is usually assumed that, as in the case with independent explanatory variables, the estimation of the n regression equations as one equation is more efficient than the estimation of each of the n regression equations separately.

For the estimator $\hat{\beta}$, it can be shown that its variance–covariance matrix is

$$\mathrm{Var}(\hat{\beta}) = (X'[\Sigma^{-1} \otimes I_N]X)^{-1} \tag{4.13}$$

and for the practical estimator $\tilde{\beta}$ its variance–covariance matrix may be taken as given by

$$\mathrm{Var}(\tilde{\beta}) = (X'[\hat{\Sigma}^{-1} \otimes I_N]X)^{-1} \tag{4.14}$$

As the matrix Σ is generally unknown for real situations, the equations (4.12) and (4.14) are more relevant than (4.11) and (4.13).

Evaluating the conditional expectations In order to be able to compute the optimal control solution to be obtained in Sections 4.3 and 4.4, it will be necessary to evaluate the expectations $E(B'H_tA)$, $E(B'H_tB)$, $E(A'H_tA)$, $E(A'H_tb)$ and $E(B'H_tb)$, which are functions of the second moments of the elements of A, B and b. Putting this evaluation problem into a more compact form, we let

$$\Pi = (A \mid B \mid b) \tag{4.15}$$

then the required expectations are given by the submatrices of

$$E(\Pi' H_t \Pi) \tag{4.16}$$

Supposing that the matrix Π has s columns which are denoted by

$\pi_1, \ldots, \pi_s$, then form the column vector π by 'stacking' the columns of Π, i.e.

$$\pi = \begin{bmatrix} \pi_1 \\ \pi_2 \\ \vdots \\ \pi_s \end{bmatrix}$$

Rearranging the elements of a matrix into a vector in this way facilitates the computation of the covariances of the elements of the matrix (Nissen 1968). If the mean and covariance matrix of the vector π are denoted by $\bar{\pi}$ and Q respectively, then we have the identity

$$\begin{aligned} E(\pi\pi') &= \bar{\pi}\bar{\pi}' + Q \\ &= \begin{bmatrix} \bar{\pi}_1\bar{\pi}_1' & \cdots & \bar{\pi}_1\bar{\pi}_s' \\ \vdots & & \vdots \\ \bar{\pi}_s\bar{\pi}_1' & \cdots & \bar{\pi}_s\bar{\pi}_s' \end{bmatrix} + \begin{bmatrix} Q_{11} & \cdots & Q_{1s} \\ \vdots & & \vdots \\ Q_{s1} & \cdots & Q_{ss} \end{bmatrix} \end{aligned}$$

Now, the $(i - j)$th element of the matrix (4.16) can then be readily computed by the following relationship (Chow 1975):

$$\begin{aligned} E(\Pi' H_t \Pi)_{ij} &= E(\pi_i' H_t \pi_j) \\ &= E \operatorname{tr} (H_t \pi_j \pi_i') \\ &= \operatorname{tr} H_t E(\pi_j \pi_i') \\ &= \operatorname{tr} H_t (\bar{\pi}_j \bar{\pi}_i' + Q_{ji}) \\ &= \bar{\pi}_i' H_t \bar{\pi}_j + \operatorname{tr} H_t Q_{ji} \end{aligned} \tag{4.17}$$

where tr denotes the trace of a matrix, which is defined as the arithmetic sum of all the elements in its leading diagonal. Thus, the required expectations in (4.16) are determined according to (4.17) once the mean vector and covariance matrix of π are known.

Once $\tilde{\beta}$ and $\mathrm{Var}(\tilde{\beta})$ are evaluated, they can be easily transformed into the form appropriate for use in the optimal control solution. For example, the estimate $\tilde{\beta}$ will furnish estimates $\tilde{A}_r$, $\tilde{B}_s$ and $\tilde{b}$ in (4.4) which will then yield $\tilde{A}$, $\tilde{B}$ and $\tilde{b}$, i.e. $\tilde{\Pi}$ defined in (4.15). Notice that in obtaining the estimates of A, B and b, only the first row elements need be estimated as the other entries are constants which have the value zero or one. Similarly, knowledge of $\mathrm{Var}(\tilde{\beta})$ will yield $\tilde{Q}$, an estimate of the variance–covariance matrix of π. We can then replace $\bar{\pi}$ and Q in

(4.17) by $\tilde{\pi}$ and $\tilde{Q}$ respectively, to give an approximate solution to the evaluation of the conditional expectations that are needed for the optimal control solution.

The Zellner approach of estimating β given in (4.7) by treating its n regression equations as one system has been applied to study the geographical dispersion of foreign workers in the Netherlands (Hordijk & Nijkamp 1977). At a more theoretical level, the approach provides an improved alternative to the spatial autocorrelation problem (Arora & Brown 1977, Bennett 1979, Paelinck & Klaassen 1979). It avoids the difficulties involved in specifying the weighting matrix associated with the spatial autoregressive disturbances suggested by Cliff and Ord (1973) since no assumption about the form of spatial autocorrelation is required. Zellner's approach does assume that the disturbances are not serially autocorrelated, but when this assumption does not hold the method has been extended by Parks (1967). He allows the errors to be both serially (from one observation to the next) and contemporaneously (from one equation to the next) correlated and assumes that the serial correlation is a first-order autoregressive process. Parks' method of estimation is again asymptotically more efficient than the equation by equation method. This result has been demonstrated to be valid even in small samples (Kmenta & Gilbert 1970).

4.3 *Estimation and control with equality constraints*

The optimal control problem considered here may be stated as follows:

Given the linear dynamic model

$$Y_t = A_t Y_{t-1} + B_t x_t + b_t + u_t \tag{2.30}$$

with the initial condition Y_0, *the optimal control problem is to determine the set of control variables* $x_1, \ldots, x_T$ *to minimise the quadratic objective function* (2.31) *when the control variables are subject to a set of k independent linear constraints* (3.1). *The parameters* A_t, B_t *and* b_t *are unknown and are assumed to be constant coefficients. The random error term* u_t *is assumed to be distributed with zero mean and covariance matrix*

$$\mathrm{Cov}(u_t, u_s) = \begin{cases} R_u & \text{if } t = s \\ 0 & \text{if } t \neq s \end{cases}$$

It is further assumed that u_t is statistically independent of A_t, B_t and b_t and that the unknown parameters are random variables. With this assumption, the following further assumption is necessary in order that practical control solutions can be obtained:

ASSUMPTION 1. *The joint probability distribution of these random variables A_t, B_t and b_t are assumed to be given at the beginning of the planning horizon and to remain constant throughout. Hence they may be written as A, B and b.*

This assumption is consistent with the viewpoint of either Bayesian or classical statistics. The Bayesian statistician will assume that the unknown parameters are random variables with known prior probability distributions derived, for example, from the least-squares estimates discussed above. On the other hand, the classical statistician expresses uncertainty in the parameters by the variances of the estimates of these parameters.

The assumption that the unknown parameters are random variables allows conditional expectations like $E_{t-1}(A_t'H_tB_t)$ at period $t - 1$ to be given by $E_0(A'H_tB)$ at period 0 and it is this simplification that enables the linear feedback control solutions to be obtained.

As in earlier chapters, the method of dynamic programming is used to solve the above problem. For the T-stage problem, dynamic programming yields a solution to the problem for the last stage T first, given all the information up to the end of the $T - 1$ previous period. Hence, as before, it is required to minimise

$$\begin{aligned} V_T &= E_{t-1}(Y_T - a_T)'M_T(Y_T - a_T) \\ &= E_{T-1}(Y_T'H_TY_T - 2Y_T'h_T + d_T) \end{aligned} \tag{4.18}$$

subject to the set of linear equality constraints

$$F_Tx_T = f_T \tag{4.19}$$

where, as previously,

$$H_T = M_T, \qquad h_T = M_Ta_T, \qquad d_T = a_T'M_Ta_T \tag{4.20}$$

Proceeding as in Chapters 2 and 3, we have the optimal control for the tth period given by

$$x_t^* = \hat{G}_tY_{t-1} + \hat{g}_t, \qquad t = 1, \ldots, T \tag{4.21}$$

where

$$\hat{G}_t = G_t - D_t^{-1}F_t'P_t^{-1}F_tG_t \tag{4.22}$$

$$\hat{g}_t = g_t - D_t^{-1}F_t'P_t^{-1}F_tg_t + D_t^{-1}F_t'P_t^{-1}f_t \tag{4.23}$$

$$G_t = -D_t^{-1}E(B'H_tA) \tag{4.24}$$

$$g_t = D_t^{-1}[E(B')h_t - E(B'H_tb)] \tag{4.25}$$

$$D_t = E(B'H_tB) \tag{4.26}$$

$$P_t = F_tD_t^{-1}F_t' \tag{4.27}$$

and H_t, h_t satisfy the following matrix Ricatti equations:

$$\begin{aligned} H_T &= M_T \\ H_{t-1} &= M_{t-1} + E(A'H_tA) + E(A'H_tB)\hat{G}_t \end{aligned} \tag{4.28}$$

$$\begin{aligned} h_T &= M_Ta_T \\ h_{t-1} &= M_{t-1}a_{t-1} + E(A')h_t - E(A'H_tB)\hat{g}_t - E(A'H_tb) \end{aligned} \tag{4.29}$$

Note that, in the above Equations (4.22)–(4.29), the time subscript t is omitted from the expectation operator and from the parameters A, B and b. This follows from Assumption 1 regarding the probability distributions of the parameters, and the fact that we are limiting our discussion to the case of time-invariant parameters. The expectations involved are functions of the second moments of the elements of A and B. Assuming that these can be evaluated, the optimal control policy (4.21) is the familiar linear feedback control rule.

The associated minimum expected loss for the T-period optimisation problem may be readily seen to be given by substituting (4.21) into (4.18) with the subscript t replaced by 1:

$$\begin{aligned} V_1^* = {} & Y_0'E[(A + B\hat{G}_1)'H_1(A + B\hat{G}_1)]Y_0 \\ & + 2Y_0'E[(A + B\hat{G}_1)'(H_1B\hat{g}_1 + H_1b - h_1)] \\ & + E[(B\hat{g}_1 + b)'H_1(B\hat{g}_1 + b)] - 2E[(B\hat{g}_1 + b)'h_1] \\ & + E(d_1) + E(u_1'H_1u_1) \end{aligned} \tag{4.30}$$

where d_1 is obtained by the following backward recursion formula,

$$\begin{aligned} d_{t-1} = {} & a_{t-1}'M_{t-1}a_{t-1} + E[(B\hat{g}_t + b)'H_t(B\hat{g}_t + b)] \\ & -2E[(B\hat{g}_t + b)'h_t] + E(u_t'H_tu_t) + E(d_t) \end{aligned} \tag{4.31}$$

given the initial condition $d_T = a_T'M_Ta_T$.

Looking at the problem from the point of view of classical statistics, the solutions (4.21) and (4.31) are still valid except that the matrices A and B should be replaced by $\tilde{A}$ and $\tilde{B}$, which are random estimators of the unknown parameters. These estimators can usually be derived by the method of least squares, as discussed in Section 4.2.

4.4 *Estimation and control with non-negativity constraints*

In the preceding section, under the specific set of assumptions made, the optimal control solution has been derived for the case in which there are constant but unknown model parameters and linear equality constraints on the control variables. This solution is no longer valid when non-negativity constraints on the control variables, such as (3.30b), are imposed. In this case it becomes necessary and, because of the computational complexity, reasonable to seek feasible solutions that are not truly optimal but only near optimal.

The presence of uncertainty in model parameters raises enormous difficulties in finding feasible control solutions subject to non-negativity control constraints for the multistage problem. Moreover, this is a control problem that has not been previously successfully solved. Although it may be possible to calculate the solution path $x_1^\circ, \ldots, x_T^\circ$ subject to only linear equality control constraints (3.30a) by setting the random terms u_t to their expected value, these control variables may not be simultaneously adjusted to satisfy the non-negativity control constraints in the usual manner. This is because a matrix, equivalent to that of $\bar{B}$ necessary for the dual simplex adjustment, is not available when the model parameters are unknown.

To overcome this difficulty, an approach is developed below which approximates the multistage solution by a step by step single-stage optimisation solution. The problem can be stated as follows:

At time $t = 1$ with Y_0 given, find the control x_1, that minimises

$$E(Y_1 - a_1)'M_1(Y_1 - a_1) \tag{4.32}$$

subject to the following constraints:

$$F_1x_1 = f_1 \tag{4.33a}$$

$$x_1 > 0 \tag{4.33b}$$

First note that, ignoring the constraint (4.33b), the solution to the above problem is given in Section 4.3 with $T = 1$. In other words, the optimal control x_1 is

$$x_1^\circ = \hat{G}_1^\circ Y_0 + \hat{g}_1^\circ \tag{4.34}$$

where

$$\hat{G}_1^\circ = G_1^\circ - \hat{D}_1^{-1}F_1'\hat{P}_1^{-1}F_1G_1^\circ \tag{4.35}$$

$$\hat{g}_1^\circ = g_1^\circ - \hat{D}_1^{-1}F_1'\hat{P}_1^{-1}F_1g_1^\circ + \hat{D}_1^{-1}F_1'\hat{P}_1^{-1}f_1 \tag{4.36}$$

$$G_1^\circ = -\hat{D}_1^{-1}E(B'M_1A) \tag{4.37}$$

$$g_1^\circ = \hat{D}_1^{-1}[E(B')M_1a - E(B'M_1b)] \tag{4.38}$$

$$\hat{D}_1 = E(B'M_1B) \tag{4.39}$$

$$\hat{P}_1 = F_1\hat{D}_1^{-1}F_1' \tag{4.40}$$

If the solution x_1° given according to (4.34) does not satisfy (4.33b), then it must be adjusted. This single-stage approach to the solution permits the dual simplex method to be used. The $\tilde{B}$ matrix of Chapter 3 is simply $E(B'M_1B) = \hat{D}_1$. Hence, with appropriate modifications to (3.37), the 'linear programming' problem can be solved to yield the adjusted solution x_1^a.

After x_1^a has been applied and Y_1 becomes known, the single-stage control problem at time $t = 2$ is given by

$$\text{minimise } E(Y_2 - a_2)'M_2(Y_2 - a_2) \tag{4.41}$$

with respect to x_2 subject to

$$F_2x_2 = f_2 \tag{4.42a}$$

$$x_2 > 0 \tag{4.42b}$$

As before, we proceed to obtain x_2^a. In general, at time t, and given all information up until that time, the single-stage control problem is

$$\text{minimise } E(Y_t - a_t)'M_t(Y_t - a_t) \tag{4.43}$$

with respect to x_t subject to

$$F_tx_t = f_t \tag{4.44a}$$

$$x_t > 0 \tag{4.44b}$$

The optimal control with just (4.44a) present is found according to

$$x_t^\circ = \hat{G}_t^\circ Y_{t-1} + \hat{g}_t^\circ \tag{4.45}$$

where

$$\hat{G}_t^\circ = G_t^\circ - \hat{D}_t^{-1} F_t' \hat{P}_t^{-1} F_t G_t^\circ \tag{4.46a}$$

$$\hat{g}_t^\circ = g_t^\circ - \hat{D}_t^{-1} F_t' \hat{P}_t^{-1} F_t g_t^\circ + \hat{D}_t^{-1} F_t' \hat{P}_t^{-1} f_t \tag{4.46b}$$

$$G_t^\circ = -\hat{D}_t^{-1} E(B' M_t A) \tag{4.46c}$$

$$g_t^\circ = \hat{D}_t^{-1}[E(B')M_t a_t - E(B' M_t b_t)] \tag{4.46d}$$

$$\hat{D}_t = E(B' M_t B) \tag{4.46e}$$

$$\hat{P}_t = F_t \hat{D}_t^{-1} F_t' \tag{4.46f}$$

For the adjustment step, replace the $\tilde{B}$ matrix of Chapter 3 by $\hat{D}_t$.

If required, this simplified method may also be used to yield the sequence of control settings from period 1 to period T at the beginning of the planning horizon. In that case it becomes necessary to replace the as yet unknown state variables $Y_1, \ldots, Y_T$ by their expected values. These expected values are readily obtained from the linear dynamical model when the appropriate control setting is applied.

The idea of reducing the original multistage problem to one that is solved as a sequence of single-stage problems is appealing as it avoids the difficulties associated with the solution of two-point boundary-value problems (Lee 1964). Naturally the solution is not optimal with respect to the multistage formulation. The main advantage of the method described lies in its computational simplicity. As the truly optimal solution to the multistage problem is not available, it is not possible to evaluate how well the solution obtained by this method approximates the true solution. Nonetheless, if the multistage problem is replaced by a sequence of single-stage problems, then the solution method of this section yields the optimal solution at each stage of the sequence of one-stage problems.

A criticism of this method is that in finding the sequence of single-stage control solutions, the future outcomes are not taken into consideration. If this feature is desired, then the following method is suggested, which is based on the discussion in Section 4.3 as well as the adjustment mechanism described in Section 3.3. For deterministic control problems the adjustment mechanism based on the dual simplex algorithm will give the truly optimal solution. At each iteration of the algorithm the basic variable with the largest negative value is made non-basic (i.e. it is made to take on the value of zero). When this basic variable is a control variable then making it non-basic amounts to adding another control constraint, namely that this control variable is equal to zero. Thus, with the current set of linear equality constraints, the feedback solution of Section 3.2 may be employed. Again, if the resulting solution has negative control variables, then we add another constraint which sets the most negative control variable to zero. The control problem with this new set of control constraints is solved again. The process is repeated until all the control variables are non-negative. It is not necessarily true that the control solution obtained in this way is the true optimum. As noted in Chapter 3, this is because of two reasons: first, the set of added control constraints at the zero level may not be the correct one; secondly, once a control variable is made to be zero, it remains so throughout the analysis, whereas in the dual simplex algorithm this is not necessarily true. However, we can take advantage of this observation in devising a method to solve the control problem considered here but with the consideration of the future outcomes.

As before, solve the control problem with only the linear equality constraint present, as discussed in Section 4.3. Although x_1^* given according to (4.21) is the only control setting to be determined at period 1, the other control settings, $x_2^*, \ldots, x_T^*$, may be replaced by their expected values by letting the random disturbance term take on its mean value. Now, when $x_1^*, \ldots, x_T^*$ are found, suppose that the ith component of x_t^* is the largest negative element. Then we add the constraint $x_{it} = 0$ to the set of linear equality constraint at period t. Thus the matrix F_t has one extra row. Using the method of Section 4.3, the problem with the further constraint $x_{it} = 0$ is solved. Again, determine $x_1^*, \ldots, x_T^*$ and, if there exist any negative control variables, repeat the above step by making the largest negative element zero. Note that at each step, only one control constraint at the zero level is added and the problem is resolved. When $x_1^*, \ldots, x_T^*$ does not have any negative components then stop and we have found the solution. Clearly, the solution obtained by this method of imposing constraints takes into account the multistage nature of the problem and will be different from the solution obtained by the sequence of single-stage problems. Of course the latter is most appealing from the computational viewpoint and this should be of prime importance in deciding on what solution method to use, unless a complete solution is found yielding the true optimal control rule.

4.5 An example

The following simple example of the control of a two-region economy will illustrate the various calculation steps necessary to determine the control rules discussed in this chapter. Assume that the state and control variables are governed by

$$Y_t = AY_{t-1} + Bx_t + u_t$$

where

$$A = \begin{bmatrix} a_{11} & a_{12} \\ a_{21} & a_{22} \end{bmatrix}, \qquad B = \begin{bmatrix} b_{11} & b_{12} \\ b_{21} & b_{22} \end{bmatrix}$$

and the error term u_t has mean zero.

Like the example in Section 3.3, the control time period is taken to be $T = 2$ and the desired target levels for the state variables are $a_1' = (5, 3)$ and $a_2' = (8, 6)$ respectively. The initial state vector is $Y_0' = (3, -1)$; the linear equality constraint matrices are $F_1 = F_2 = (1, 1)$, while $f_1 = 5$ and $f_2 = 8$. The weighting matrices M_1 and M_2 in the objective function are taken to be the identity matrix.

We assume that the parameters A and B have been estimated. Following the notation in Section 4.2 we assume that

$$\bar{\pi}' = (1.6, -0.2, 0.8, 1.8, 0.7, 1.8, -0.1, 0.9)$$

and for further simplification let us suppose that the covariance matrix Q is diagonal and is given by

$$Q = \text{diag}(0.05, 0.05, 0.05, 0.05, 0.05, 0.05, 0.05, 0.05)$$

For the problem with only linear equality constraints present, we have to determine the optimal controls x_1^* and x_2^* as given by (4.19). From the initial conditions

$$H_2 = \begin{bmatrix} 1 & 0 \\ 0 & 1 \end{bmatrix} \quad \text{and} \quad h_2 = \begin{bmatrix} 8 \\ 6 \end{bmatrix}$$

together with $\bar{\pi}$, Q and (4.15) we obtain

$$E(A'H_2A) = \begin{bmatrix} 2.70 & 0.92 \\ 0.92 & 3.98 \end{bmatrix}, \quad E(B'H_2B) = \begin{bmatrix} 3.83 & 1.55 \\ 1.55 & 0.92 \end{bmatrix},$$

$$E(A'H_2B) = \begin{bmatrix} 0.76 & -0.34 \\ 3.80 & 1.54 \end{bmatrix}$$

These conditional expectations enable the calculation of the gain matrix $\hat{G}_2$ and the forcing vector $\hat{g}_2$. According to (4.20) and (4.21) we have

$$\hat{G}_2 = \frac{1}{1.8498}\begin{bmatrix} -1.2332 & -2.5337 \\ 1.2332 & 2.5337 \end{bmatrix}$$

and

$$\hat{g}_2 = \frac{1}{1.8498}\begin{bmatrix} -22.2426 \\ 37.0411 \end{bmatrix}$$

Then from the recursion equations for H and h, (4.26) and (4.27), we obtain

$$H_1 = \frac{1}{1.8498}\begin{bmatrix} 5.4878 & -1.0852 \\ -1.0852 & 3.4859 \end{bmatrix}$$

and

$$h_1 = \frac{1}{1.8498}\begin{bmatrix} 71.3042 \\ 50.0463 \end{bmatrix}$$

Again, together with $\bar{\pi}$, Q and (4.15), the conditional expectations are determined as follows:

$$E(A'H_1A) = \begin{bmatrix} 8.2881 & 1.5232 \\ 1.5232 & 6.5574 \end{bmatrix}, \quad E(B'H_1B) = \begin{bmatrix} 6.3236 & 2.5812 \\ 2.5812 & 1.9043 \end{bmatrix},$$

$$E(A'H_1B) = \begin{bmatrix} 1.0368 & -1.6704 \\ 6.1831 & 2.4987 \end{bmatrix}$$

From (4.20) and (4.21) it can be checked that

$$\hat{G}_1 = \frac{1}{16.4895}\begin{bmatrix} -14.5625 & -19.8187 \\ 14.5625 & 19.8187 \end{bmatrix}$$

and

$$\hat{g}_2 = \frac{1}{16.4895}\begin{bmatrix} -391.8222 \\ 474.2697 \end{bmatrix}$$

Hence, according to (4.19), the optimal control for the first period when only the constraint $x_{1,1} + x_{2,1} = 5$ is present is given by

$$x_1^* = \begin{bmatrix} -25.2094 \\ 30.2094 \end{bmatrix}$$

Setting the error term u to its mean value, i.e. zero, gives $Y_1' = (-16.6675, -20.5885)$ so that, from the linear feedback equation (4.19),

$$x_2^* = \begin{bmatrix} 27.2877 \\ -19.2877 \end{bmatrix}$$

This satisfies the linear equality constraint at the second time period.

For the case when non-negativity constraints on the control variables at each time period are added, the solution strategies presented in Section 4.4 give the same control solutions, namely

$$\begin{bmatrix} 0 \\ 5 \end{bmatrix} \text{ and } \begin{bmatrix} 0 \\ 8 \end{bmatrix}$$

at $t = 1$ and $t = 2$ respectively. The solution strategy using the method of imposing constraints yields the above non-negative control solutions more readily than the other strategy of solving a step by step single-stage problem.

4.6 Conclusion

In this chapter, a strategy has been given for solving the control problem with constant but unknown parameters and subject to linear equality and non-negativity constraints. The strategy is in two parts. First, the unknown parameters are estimated by the method of least squares. This allows evaluation of the various conditional expectations that are required. Secondly, the control policy is determined. Note that the estimates can be held constant or can be updated after the model has been subject to control when more information becomes available. Conceptually, re-estimating the parameters when more data becomes available will yield better control, but at the expense of more computational effort. However, when the control period is considerably shorter than the estimation period, there is usually not much to be gained by updating the estimates (see, for example, Chow 1975).

Connected to the idea of updating the estimates of the unknown parameters during the control period is the method of the self-tuning regulator proposed for systems with single input and output (Åstrom &

Wittenmark 1973). The self-tuning regulator is a sub-optimal controller and is obtained in two parts. First, the unknown parameters are determined by least-squares estimation. Secondly, these estimates are substituted into the control law that is derived assuming that the parameters were completely known. These two parts of the control strategy are repeated at each time period. In other words, the parameter estimates are revised as more information becomes available. Self-tuning regulators are asymptotically optimal and converge to the minimum variance controllers. However, general results giving convergence conditions for controllers which have the self-tuning property are not yet available, especially for the multivariate case. In the meantime, the control solution suggested here yields a good practical means of calculating sub-optimal solutions in conditions of controlling unknown systems subject to linear equality and non-negativity constraints. In the next chapter, we extend the analysis to systems with unknown and time-varying parameters.

5 *Spatial control with time-varying parameters*

5.1 Time-varying parameters

In this chapter, we extend the analysis of the preceding chapters by allowing the model parameters to be not only unknown but also time-varying. This is accomplished by a form of Kalman filter similar to the Åstrom and Wittenmark self-tuning regulator. Although numerical results, for example by Chow (1975), indicate that in many applications there is little to be gained by updating the estimates of the unknown but constant model parameters during the control period, this conclusion does not usually apply to the more general case of unknown *time-varying* parameters. Conceptually, this is because, when the parameters are time-varying, control settings that are based on revised parameter estimates reflecting the changing nature of these unknown parameters are more accurate than control settings that are based on parameter estimates which are not updated. The relative importance of updating depends on the speed of parameter variation. If the unknown parameters are changing rapidly, then it is essential that rapid updating of their estimates be incorporated in the design of the control rule. However, for this case it is at present only possible to obtain control solutions that are approximately optimal.

Considerable attention has been given to the statistical problem of estimating time-varying parameters in linear regression systems (Farison 1967, Burnett & Guthrie 1970, Duncan & Horn 1972; Rosenberg 1972, Brown *et al*. 1975, Bennett 1979, Ch. 5). The need to control the linear system is absent from most of these analyses. The joint problem of estimation and control of linear system with time-varying parameters is not an easy one and the results that have been obtained to date are for systems without constraints on the control variables. For example, Åstrom and Wittenmark (1971) consider the problem for linear systems under control but with only single input and output and no constraints, whilst Farison *et al*. (1967) impose a method to control the usual linear dynamical system but without any control constraints such that the estimation and control portions of the problem can be performed separately. This chapter extends the approach of Farison *et al*. to the situation where there are constraints on the control variables.

It is convenient to define the problem formally. The linear dynamical model is represented in the familiar form:

$$Y_t = A_t Y_{t-1} + B_t x_t + b_t + u_t \tag{5.1}$$

with given initial condition $Y_0 = \bar{Y}_0$.

The optimal control problem is to determine the set of control variables $x_1, \ldots, x_T$ to minimise a quadratic function (the specific form of which will be stated later) when the control variables are subject to a set of k independent linear equality constraints

$$F_t x_t = f_t, \qquad t = 1, \ldots, T \tag{5.2a}$$

and the non-negativity restrictions

$$x_t \geq 0, \qquad t = 1, \ldots, T \tag{5.2b}$$

The parameters A_t, B_t and b_t in (5.1) are unknown and are assumed to be time-varying. A useful assumption which is made in the following is that the changing nature of the parameter is governed by a Gauss–Markov process. This assumption is not essential but simplifies the discussion. Other possible assumptions are discussed by Bennett (1979, Ch. 5). Collecting the parameters into a vector of appropriate dimension ϕ_t, the Gauss–Markov process is described by the stochastic difference equation

$$\phi_t = C_t \phi_{t-1} + v_t \tag{5.3}$$

where C_t is a known constant matrix and v_t is a random vector. Note that the case of unknown constant parameters is included in (5.3) when $C_t = I$ and $v_t = 0$.

It is assumed that $\{(u_t, v_t), t = 1, 2, \ldots\}$ is a sequence of independent identically distributed random vectors which are normally distributed with zero means and known covariance

$$\begin{pmatrix} R_u & 0 \\ 0 & R_v \end{pmatrix}$$

Furthermore, u_t is independent of ϕ_t and $Y_{t-1}, Y_{t-2}, \ldots, x_t, x_{t-1}, \ldots, b_t, b_{t-1}, \ldots$ while v_t is independent of $\phi_{t-1}, \phi_{t-2}, \ldots$.

Without the constraints (5.2), the method of Farison *et al.* (1967) is based upon using a conditional quadratic cost objective function which permits separation of the estimation and control aspects of the problem. Briefly, the basic idea of the method is as follows. At the current time period t, the unknown parameters are estimated. With this as the 'present' model of the system the control for time t is determined. Upon applying this control and measuring the new information, the parameters are re-estimated giving the model of the system. The control for this time period $t + 1$ is then evaluated and the cycle repeated. The algorithms for the two portions of the problem may be easily implemented in machine calculation and for this reason this approach allows extension to the problem with constraints. It turns out that the estimation part of the problem is unaffected by the presence of the constraints on the control variables, but the control part of the problem must be modified.

5.2 *Estimation of the time-varying parameters*

The parameter vector ϕ_t is obtained from the system parameters (A_t, B_t and b_t) and in the following it will be convenient to re-express (5.1) in terms of ϕ_t. A simple example indicating how this is achieved is now considered. Ignoring the random term u_t, assume that (5.1) is given as

$$\begin{bmatrix} Y_{1t} \\ Y_{2t} \end{bmatrix} = \begin{bmatrix} a_{11,t} & a_{12,t} \\ a_{21,t} & a_{22,t} \end{bmatrix} \begin{bmatrix} Y_{1,t-1} \\ Y_{2,t-1} \end{bmatrix} + \begin{bmatrix} b_{11,t} & b_{12,t} \\ b_{21,t} & b_{22,t} \end{bmatrix} \begin{bmatrix} x_{1,t} \\ x_{2,t} \end{bmatrix} + \begin{bmatrix} b_{1,t} \\ b_{2,t} \end{bmatrix} \tag{5.4a}$$

The parameters are collected into a vector form, column by column, as follows:

$$\phi_t' = (a_{11,t}, a_{21,t}, a_{12,t}, a_{22,t}, b_{11,t}, b_{21,t}, b_{12,t}, b_{22,t}, b_{1,t}, b_{2,t}).$$

Hence, (5.4a) may be rewritten as

$$\begin{aligned} \begin{bmatrix} Y_{1t} \\ Y_{2t} \end{bmatrix} &= \left[\begin{array}{cccc|cccc|cc} Y_{1,t-1} & 0 & Y_{2,t-1} & 0 & x_{1,t} & 0 & x_{2,t} & 0 & 1 & 0 \\ 0 & Y_{1,t-1} & 0 & Y_{2,t-1} & 0 & x_{1,t} & 0 & x_{2t} & 0 & 1 \end{array}\right] \phi_t \\ &= S_t \phi_t \end{aligned} \tag{5.4b}$$

where the definition of the matrix S_t is clear. This shows how to re-express (5.1) in terms of ϕ_t. Thus, with matrix S_t of suitable dimension,

$$Y_t = S_t \phi_t + u_t \tag{5.5}$$

where the (i, j)th element of S_t is the coefficient of the jth component of ϕ_t in the ith row of (5.1). Note that S_t depends only on Y and x.

The first objective is to estimate the unknown parameters ϕ_t. It turns out that this estimation can be accomplished recursively via the well known Kalman filter equation. This becomes evident immediately if we note that the relevant equations

$$\phi_t = C_t\phi_{t-1} + v_t \tag{5.6}$$

$$Y_t = S_t\phi_t + u_t \tag{5.7}$$

are similar to the unconstrained control problem with incomplete state information of Section 2.5. Specifically, (5.6), which describes the evolution of the unknown parameters, is now the state equation of the system, while (5.7), which has previously been the state equation, now becomes the measurement equation. Of course the state of the present system is ϕ_t, which is not directly measurable but can only be observed through Y_t.

The normality assumptions of the random terms v_t and u_t imply that we only need to obtain the conditional mean and covariance matrix of the unknown parameters. These may also be obtained recursively. Keeping to the same kind of notation as in Section 2.5 we let

$$E(\phi_t/Y_{t-1}) = \phi_{t/t-1}$$
$$E(\phi_t - \phi_{t/t-1})(\phi_t - \phi_{t/t-1})' = \Gamma_{t/t-1}.$$

Then the Kalman filter equations are given by

$$\phi_{t/t-1} = C_t\phi_{t-1/t-1} \tag{5.8a}$$

$$\Gamma_{t/t-1} = C_t\Gamma_{t-1/t-1}C_t' + R_v \tag{5.8b}$$

$$\Phi_t = \Gamma_{t/t-1}S_t'[S_t\Gamma_{t/t-1}S_t' + R_u]^{-1} \tag{5.8c}$$

$$\phi_{t/t} = \phi_{t/t-1} + \Phi_t(Y_t - S_t\phi_{t/t-1}) \tag{5.8d}$$

$$\Gamma_{t/t} = \Gamma_{t/t-1} - \Phi_t S_t\Gamma_{t/t-1} \tag{5.8e}$$

with given initial conditions

$$\phi_{0/0} = \bar{\phi}_0 \tag{5.9a}$$

$$\Gamma_{0/0} = \Gamma_0 \tag{5.9b}$$

The equations (5.8) specify one complete step in the estimation process, with (5.8d) and (5.8e) representing the updating equations when Y_t becomes known. Equation (5.8d) shows the recursive nature of the computations. The optimal estimator of ϕ_t at time t, $\phi_{t/t}$, is given as a linear combination of the previous estimator and of the current observation Y_t.

There are difficulties in using the above equations if the error covariance matrices R_u and R_v are unknown. These can be estimated, but it is more usual to invoke some assumed structure to ensure that the Kalman gain matrix Φ_t does not vanish as t increases; otherwise from (5.8d) and (5.8e) there will be no change in the mean and covariance matrix of the parameters (a phenomenon termed *divergence*). Assumptions have to be made or some *ad hoc* methods must be used to prevent Φ_t from vanishing, thereby keeping the Kalman filter equations open to new information (see Bennett 1979, Ch. 5).

At time t, when Y_t is known and $\phi_{t/t}$ and $\Gamma_{t/t}$ evaluated, the control portion of the problem is now considered. The evaluation of the control setting to be applied requires estimates of the future values of ϕ and its covariance matrix. For any $j > t$, these estimates may be calculated from (see Farison *et al.* 1967)

$$\phi_{j/t} = C(j, t)\phi_{t/t} \tag{5.10}$$

$$\Gamma_{j/t} = C(j, t)\Gamma_{t/t}C'(j, t) + \sum_{i=t}^{j-1} C(j, i+1)R_v C'(j, i+1) \tag{5.11}$$

where

$$C(j, t) = C_j C_{j-1} \ldots C_{t+1} \qquad \text{for } j > t$$

$$C(t, t) = I$$

As shown in the next section, where only linear equality control constraints are imposed, the estimating equations (5.10) and (5.11) are necessary in order that the optimal control rule at time t be determined. These two equations take into consideration the time-varying nature of the model parameters according to (5.3).

5.3 *Control of equality-constrained time-varying systems*

To achieve separation of the estimation and control parts of the problem, following Farison *et al.* (1967), the objective function can be defined to be the conditional quadratic loss function. Specifically, at time k ($k = 0, \ldots, T-1$), and with information available only up to time k, the objective function for the remaining $T - k$ stages is

$$V(T-k/k) = E\left[\sum_{t=k+1}^{T} (Y_t - a_t)'M_t(Y_t - a_t)/k\right] \tag{5.12}$$

The conditional nature of the objective function is indicated by $/k$. This states that the control problem at time k, with $T - k$ stages to go, is based on the current model of the dynamical system which has just been estimated. Subject only to the linear equality constraints (5.2a), the control problem at time k is to find the sequence of control inputs $x(t/k), t = k + 1, \ldots, T$ such that (5.12) is minimised. After the control $x(k + 1/k)$ is applied, Y_{k+1} becomes available and the system model can be updated using the Kalman filter of Section 5.3. Then, at time $k + 1$, the procedure is repeated. In effect the overall control problem is to find the control inputs $x(k + 1/k)$ for each $k = 0, \ldots, T - 1$. This approach for the control portion of the problem does take into account the multistage nature of the problem. Furthermore, at each time period the control inputs derived are based on all information available at that time. As expected, this strategy needs to be modified when inequality constraints (5.2b) are added. This problem is considered in the next section.

The determination of $x(j/k), j = k + 1, \ldots, T$ at time k to minimise (5.12) subject to the linear equality constraints $F_t x_t = f_t$ for $t = k + 1, \ldots, T$ may be performed in the usual manner as in Section 4.3, using the method of dynamic programming with Lagrange multipliers. Thus for the $(T - k)$-stage control problem, the method yields the control setting for the final stage (i.e. T) first, by minimising the function

$$\begin{aligned} V_T(T-k/k) &= E\{(Y_T - a_T)'M_T(Y_T - a_T)/k\} \\ &= E_k(Y_T - a_T)'M_T(Y_T - a_T) \end{aligned} \tag{5.13}$$

subject to the constraint

$$F_T x_T = f_T \tag{5.14}$$

where E_k in (5.13) denotes expectation conditioned on the present system model at time k. The control variable satisfying (5.14) and minimising (5.13) may be readily found from minimising the Lagrangian

$$L_T = V_T(T - k/k) - 2\lambda_T'(f_T - F_T x_T) \tag{5.15}$$

with respect to x_t and λ_T, the Lagrange multipliers. Going through the same kind of calculations as in Sections 3.2 and 4.3, the optimal control variable at time T is given in the familiar linear feedback form. At the next step the method considers the control problem from period $T-1$ onwards, again based on the current system model at time k. This step is repeated until the optimal control at time $k+1$ is found. The sequence of optimal inputs $x^e(t/k)$ at time k for $t = k+1, \ldots, T$ is given by

$$x^e(t/k) = \hat{G}_{t/k} Y_{t-1} + \hat{g}_{t/k}, \qquad t = k+1, \ldots, T \tag{5.16}$$

where

$$\hat{G}_{t/k} = G_{t/k} - D_{t/k}^{-1} F_t' P_{t/k}^{-1} F_t G_{t/k} \tag{5.17a}$$

$$\hat{g}_{t/k} = g_{t/k} - D_{t/k}^{-1} F_t' P_{t/k}^{-1} F_t g_{t/k} + D_{t/k}^{-1} F_t' P_{t/k}^{-1} f_t \tag{5.17b}$$

$$G_{t/k} = -D_{t/k}^{-1} E_k(B_t' H_t A_t) \tag{5.17c}$$

$$g_{t/k} = D_{t/k}^{-1}[E_k(B_t')h_t - E_k(B_t' H_t b_t)] \tag{5.17d}$$

$$D_{t/k} = E_k(B_t' H_t B_t) \tag{5.17e}$$

$$P_{t/k} = F_t D_{t/k}^{-1} F_t' \tag{5.17f}$$

and the parameters H_t and h_t satisfy the following recursions:

$$\begin{aligned} H_T &= M_T \\ H_{t-1} &= M_{t-1} + E_k[(A_t + B_t \hat{G}_{t/k})' H_t (A_t + B_t \hat{G}_{t/k})] \\ &= M_{t-1} + E_k(A_t' H_t A) + E_k(A_t' H_t B_t) \hat{G}_{t/k} \end{aligned} \tag{5.18a}$$

$$\begin{aligned} h_T &= M_T a_T \\ h_{t-1} &= M_{t-1} a_{t-1} + E_k[(A_t + B_t \hat{G}_{t/k})'(h_t - H_t B_t \hat{g}_{t/k} - H_t b_t)] \\ &= M_{t-1} a_{t-1} + E_k(A_t') h_t - E_k(A_t' H_t B_t) \hat{g}_{t/k} - E_k(A_t' H_t b_t). \end{aligned} \tag{5.18b}$$

Clearly the setting of $x^e(k+1/k)$ given by (5.16) is dependent upon the evaluation of the various conditional expectations in (5.17) and (5.18). These expectations (e.g. $E_k(A_t' H_t A_t)$, $E_k(B_t')$, etc.) are conditional upon the current system model at time k and they can be easily obtained directly from (5.10), (5.11) and the relationship (4.15), which expresses the second moments of the unknown parameters in terms of their mean and covariances. Note that at each k, only $x^e(k+1/k)$ is actually applied. When Y_{k+1} becomes available, the system model is updated at period $k+1$ and $x^e(k+2/k+1)$ is determined by performing the necessary calculations specified by (5.16), (5.17) and (5.18), with the index k increased to $k + 1$. The updating of the system model from period k to $k + 1$ upon measurement of the new data Y_{k+1} may be carried out according to Equations (5.8d) and (5.8e).

5.4 *Control of non-negativity-constrained time-varying systems*

When inequality constraints on the controls (constraints (5.2b)) are added to the multistage control problem discussed in the preceding section it becomes impossible to use the adjustment mechanism based on the dual simplex method adopted previously. This difficulty has been encountered in the discussion of Section 4.4 and arises because the unknown model parameters make unavailable a matrix equivalent to that of $\bar{B}$ which is necessary for the adjustment mechanism to work. It turns out that the same philosophy of Section 4.4 can be applied here also, i.e. the multistage solution can be approximated by a step by step single-stage optimisation solution. Alternatively, the method of imposing constraints may be used (see Ch. 3).

At time k ($k = 0, \ldots, T - 1$), based on the current model of the system, the single-stage control problem is to select $x(k + 1/k)$ that minimises the one-stage objective function

$$V(1/k) = E[(Y_{k+1} - a_{k+1})'M_{k+1}(Y_{k+1} - a_{k+1})/k] \tag{5.19}$$

subject to the constraints at that time:

$$F_{k+1}x_{k+1} = f_{k+1} \tag{5.20a}$$

$$x_{k+1} > 0 \tag{5.20b}$$

To find the step by step single-stage solution, first consider the presence of only (5.20a). It has been shown in the previous section that in this case the optimal control $x^{e}(k + 1/k)$ is given in linear feedback form as

$$x^{e}(k + 1/k) = \hat{G}^{\circ}_{k+1/k} Y_k + \hat{g}^{\circ}_{k+1/k} \tag{5.21}$$

where

$$\hat{G}^{\circ}_{k+1/k} = G^{\circ}_{k+1/k} - \hat{D}^{-1}_{k+1/k} F'_{k+1} \hat{P}^{-1}_{k+1/k} F_{k+1} G^{\circ}_{k+1/k} \tag{5.22a}$$

$$\hat{g}^{\circ}_{k+1/k} = g^{\circ}_{k+1/k} - \hat{D}^{-1}_{k+1/k} F'_{k+1} \hat{P}^{-1}_{k+1/k} F_{k+1} g^{\circ}_{k+1/k} + \hat{D}^{-1}_{k+1/k} F'_{k+1} \hat{\mathrm{P}}^{-1}_{k+1/k} f_{k+1} \tag{5.22b}$$

$$G^{\circ}_{k+1/k} = -\hat{D}^{-1}_{k+1/k} E_k(B'_{k+1} M_{k+1} A_{k+1}) \tag{5.22c}$$

$$g^{\circ}_{k+1/k} = \hat{D}^{-1}_{k+1/k}[E_k(B'_{k+1}) M_{k+1} \mathrm{a}_{k+1} - E_k(B_{k+1} M_{k+1} b_{k+1})] \tag{5.22d}$$

$$\hat{D}_{k+1/k} = E_k(B'_{k+1} M_{k+1} B_{k+1}) \tag{5.22e}$$

$$\hat{P}_{k+1/k} = F_{k+1} \hat{D}^{-1}_{k+1/k} F'_{k+1} \tag{5.22f}$$

When $x^e(k + 1/k)$ is evaluated and it also satisfies (5.2b), then it is not necessary to adjust the solution. Otherwise, for the adjustment step the matrix $\bar{B}$ of Section 3.3 required for the dual simplex algorithm is replaced by $\hat{D}_{k+1/k}$, which is given by (5.22e). Denote the adjusted solution by $x^a(k + 1/k)$. When this is applied, and Y_{k+1} is available for measurement, the system model can be updated according to the equations in Section 5.2. Then the whole single-stage control problem is repeated at the next time period $k + 1$. The equations (5.21) and (5.22) are valid, but the index k is replaced by $k + 1$, throughout. Although this approach does not give the true optimal solution to the multistage problem, it does yield solutions that are feasible and are at least the 'single-stage optimisers'. The main advantage lies in its simplicity for real-time computation.

As in Section 4.4, the method of imposing constraints may be used as an alternative here. This alternative method does take into account the multistage nature of the problem. At time k, the $(T - k)$-stage control problem is solved subject to the linear equality control constraints (5.22a) only. This will yield, according to Section 5.3, the control setting $x^e(k + 1/k)$. On setting the random disturbance term to zero, the expected controls for time $t = k + 2, \ldots, T$ may be obtained. Then, for that control variable that is most negative, impose a constraint that sets this variable equal to zero. With this new constraint, re-solve the control problem by applying the method of Section 5.3. At each 'iteration' a new constraint which sets a control variable to zero is added and the problem solved again. This is repeated until all the control variables are non-negative.

5.5 *An example*

The following simple example of the control of a two-region economy will illustrate the various calculation steps necessary to determine the control rules discussed in this chapter. This extends the example introduced in Section 4.5 to the case of unknown time-varying parameters. Assume that the state and control variables are governed by the following system equation:

$$\begin{bmatrix} Y_{1t} \\ Y_{2t} \end{bmatrix} = \begin{bmatrix} a_{11,t} & a_{12,t} \\ a_{21,t} & a_{22,t} \end{bmatrix} \begin{bmatrix} Y_{1,t-1} \\ Y_{2,t-1} \end{bmatrix} + \begin{bmatrix} b_{11,t} & b_{12,t} \\ b_{21,t} & b_{22,t} \end{bmatrix} \begin{bmatrix} x_{1t} \\ x_{2t} \end{bmatrix} + \begin{bmatrix} u_{1t} \\ u_{2t} \end{bmatrix} \tag{5.23}$$

Keeping to the same notation as in Section 5.2, it is further assumed that the unknown time-varying model parameters collected into the vector ϕ_t satisfy the following equation:

$$\underset{(8\times 1)}{\phi_t} = \underset{(8\times 8)}{C}\,\underset{(8\times 1)}{\phi_{t-1}} + \underset{(8\times 1)}{v_t} \tag{5.24}$$

where for simplicity we take the matrix C to be unchanging and diagonal, i.e.

$$C = \text{diag}(1.1, 1.1, 1.1, 1.1, 1.1, 1.1, 1.1, 1.1).$$

Like the examples in Sections 3.3 and 4.5, the control time period is taken to be $T = 2$; the desired targets for the state variables are $a_1' = (5, 3)$ and $a_2' = (8, 6)$; the initial state vector is $Y_0' = (3, 1)$; the linear equality constraint matrices are $F_1 = F_2 = (1, 1)$, while $f_1 = 5$, $f_2 = 8$; and the weighting matrices M_1 and M_2 in the objective function are taken to be the identity matrix.

To be able to begin evaluating the optimal control rules, the covariance matrices of the random terms u and v need to be given. We assume *a priori* that $R_u = \text{diag}(0.02, 0.02)$ and $R_v = \text{diag}(0.01, 0.01, 0.01, 0.01, 0.01, 0.01, 0.01, 0.01)$. Furthermore, we assume that the mean vector and covariance matrix of ϕ at time $k = 0$ are found to be

$$\bar{\phi}_0' = (1.60, -0.20, 0.80, 1.80, 0.70, 1.80, -0.10, 0.90)$$

$$\Gamma_0 = \text{diag}(0.05, 0.05, 0.05, 0.05, 0.05, 0.05, 0.05, 0.05)$$

Equality-constrained solution We consider first the problem with only linear equality constraints present. Thus we have to determine $x^e(1/0)$ and $x^e(2/1)$ as given by (5.16). Accordingly, at time $k = 0$, it follows from (5.10) and (5.11) that

$$\phi_{1/0}' = [1.76, -0.22, 0.88, 1.98, 0.77, 1.98, -0.11, 0.99] \tag{5.25a}$$

$$\phi_{2/0}' = [1.936, -0.242, 0.968, 2.178, 0.847, 2.178, -0.121, 1.089] \tag{5.25b}$$

while the corresponding covariance matrices estimated at time $k = 0$ are

$$\Gamma'_{1/0} = \text{diag}[0.0705, 0.0705, 0.0705, 0.0705, 0.0705, 0.0705, 0.0705, 0.0705] \tag{5.26a}$$

$$\Gamma'_{2/0} = \text{diag}[0.1074, 0.1074, 0.1074, 0.1074, 0.1074, 0.1074, 0.1074, 0.1074] \tag{5.26b}$$

Then, from the initial conditions

$$H_2 = \begin{bmatrix} 1 & 0 \\ 0 & 1 \end{bmatrix} \quad \text{and} \quad h_2 = \begin{bmatrix} 8 \\ 6 \end{bmatrix}$$

together with (5.25b), (5.26b) and (4.17), we obtain

$$E_0(A'_2 H_2 A_2) = \begin{bmatrix} 4.0215 & 1.3470 \\ 1.3470 & 5.8950 \end{bmatrix},$$

$$E_0(B'_2 H_2 B_2) = \begin{bmatrix} 5.6759 & 2.2694 \\ 2.2694 & 1.4154 \end{bmatrix}$$

$$E_0(A'_2 H_2 B_2) = \begin{bmatrix} 1.1127 & -0.4978 \\ 5.5636 & 2.2547 \end{bmatrix} = E_0(B'_2 H_2 A_2)'$$

These conditional expectations enable the calculation of the gain matrix $\hat{G}_{2/0}$ and forcing vector $\hat{g}_{2/0}$. Performing the computations given in (5.17), it turns out that

$$\hat{G}_{2/0} = \frac{1}{7.3601} \begin{bmatrix} -4.6438 & -9.5411 \\ 4.6438 & 9.5411 \end{bmatrix},$$

$$\hat{g}_{2/0} = \frac{1}{7.3601} \begin{bmatrix} 21.4704 \\ 37.4107 \end{bmatrix}$$

Then from (5.18a) and (5.18b) the recursion equations for H and h, we get

$$H_1 = \frac{1}{7.3601} \begin{bmatrix} 29.4799 & -5.4518 \\ -5.4518 & 19.1811 \end{bmatrix}$$

and

$$h_1 = \frac{1}{7.3601} \begin{bmatrix} 134.8398 \\ -28.5439 \end{bmatrix}$$

Again, together with (5.25a) and (5.26a) and (4.17), the conditional expectations determined are

$$E_0(A_1'H_1A_1) = \begin{bmatrix} 13.5729 & 2.6304 \\ 2.6304 & 11.2035 \end{bmatrix},$$

$$E_0(B_1'H_1B_1) = \begin{bmatrix} 10.7992 & 4.3659 \\ 4.3659 & 3.2301 \end{bmatrix}$$

$$E_0(A_1'H_1B_1) = \begin{bmatrix} 1.8371 & -2.6516 \\ 10.5110 & 4.2367 \end{bmatrix} = E_0(B_1'H_1A_1)'$$

Finally, from (5.17) we obtain

$$\hat{G}_{1/0} = \frac{1}{83.8139}\begin{bmatrix} -71.0173 & -99.2684 \\ 71.0173 & 99.2684 \end{bmatrix}$$

$$\hat{g}_{1/0} = \frac{1}{83.8139}\begin{bmatrix} 104.4784 \\ 314.5909 \end{bmatrix}$$

The optimal control for the first period when only linear equality constraint $x_{1,1} + x_{2,1} = 5$ is present is given according to (5.16), which yields

$$x^e(1/0) = \begin{bmatrix} -0.1110 \\ 5.1110 \end{bmatrix}$$

When $x^e(1/0)$ is applied, and assuming that the random term u in (5.23) takes on its expected value (i.e. zero), then $Y_1' = (3.7523, 2.2001)$. With $\hat{G}_{2/0}$ and $\hat{g}_{2/0}$ known, it follows from the linear feedback equation (5.16) that

$$x^e(2/0) = \begin{bmatrix} -2.3024 \\ 10.3024 \end{bmatrix}$$

which clearly satisfies the linear equality constraint at the second time period. Note that $x^e(2/0)$ given above is determined at time $k = 0$ and without updating the parameter estimates.

Next, we evaluate $x^e(2/1)$ at time $k = 1$ given that $Y_1' = (3.7523, 2.2001)$ is known. First the parameter estimates must be updated according to (5.8d) and (5.8e). With

$$S_1 = \left[\begin{array}{cccc|cccc} 3 & 0 & -1 & 0 & -0.111 & 0 & 5.111 & 0 \\ 0 & 3 & 0 & -1 & 0 & -0.111 & 0 & 5.111 \end{array}\right]$$

it turns out that

$$\phi_{1/1} = \phi_{1/0}, \qquad \phi_{2/1} = \phi_{2/0}$$

but

$$\Gamma_{2/1} = \frac{1}{2.5674}\begin{bmatrix} 0.1906 & 0 & 0.018 & 0 & 0.0019 & 0 & -0.0922 & 0 \\ 0 & 0.1960 & 0 & 0.018 & 0 & 0.0019 & 0 & -0.0922 \\ 0.018 & 0 & 0.2387 & 0 & -0.0007 & 0 & 0.0307 & 0 \\ 0 & 0.018 & 0 & 0.2387 & 0 & -0.0007 & 0 & 0.0307 \\ 0.0019 & 0 & -0.0007 & 0 & 0.2446 & 0 & 0.0034 & 0 \\ 0 & 0.0019 & 0 & -0.0007 & 0 & 0.2446 & 0 & 0.0034 \\ -0.0922 & 0 & 0.0307 & 0 & 0.0034 & 0 & 0.0876 & 0 \\ 0 & -0.0922 & 0 & 0.0307 & 0 & 0.0034 & 0 & 0.0876 \end{bmatrix}$$

and using (4.17) the following equations are obtained:

$$E_1(A_2'H_2A_2) = \begin{bmatrix} 3.9552 & 1.3610 \\ 1.3610 & 5.8666 \end{bmatrix},$$

$$E_1(B_2'H_2B_2) = \begin{bmatrix} 5.6516 & 2.2720 \\ 2.2720 & 1.2688 \end{bmatrix}$$

$$E_1(A_2'H_2B_2) = \begin{bmatrix} 1.1142 & -0.5696 \\ 5.5631 & 2.2786 \end{bmatrix} = E_1(B_2'H_2A_2)'$$

Then from (5.17) we find that

$$\hat{G}_{2/1} = \frac{1}{4.7737}\begin{bmatrix} -3.3823 & -6.5979 \\ 3.3823 & 6.5979 \end{bmatrix},$$

$$\hat{g}_{2/1} = \frac{1}{4.7737}\begin{bmatrix} 12.5593 \\ 25.6304 \end{bmatrix}$$

and with $Y_1' = (3.7523, 2.2001)$, the optimal control at time $k = 1$ for the second time period is found from (5.16) to be equal to

$$x^e(2/1) = \begin{bmatrix} -3.0685 \\ 11.0685 \end{bmatrix}$$

Note the difference between $x^e(2/1)$ and $x^e(2/0)$. If these two control settings are applied and the random term in (5.23) is $u_2 = 0$, then, at the last stage of the planning horizon,

$$\begin{pmatrix}\text{loss function when} \\ x^e(2/0) \text{ is used}\end{pmatrix} - \begin{pmatrix}\text{loss function when} \\ x^e(2/1) \text{ is used}\end{pmatrix}$$
$$= 19.9652 - 17.0628 = 2.9024$$

which implies that the method with updating of the estimates gives better control performance of the system. To summarise, the solutions are as follows:

(a) *Without updating*

$$x^e(1/0) = \begin{bmatrix} -0.1110 \\ 5.1110 \end{bmatrix}, \qquad x^e(2/0) = \begin{bmatrix} -2.3024 \\ 10.3024 \end{bmatrix}$$

(b) *With updating*

$$x^e(1/0) = \begin{bmatrix} -0.1110 \\ 5.1110 \end{bmatrix}, \qquad x^e(2/1) = \begin{bmatrix} -3.0685 \\ 11.0685 \end{bmatrix}$$

Non-negativity-constrained solution When non-negativity constraints on the control variables are added, the solution strategy calls for two single-stage control problems to be solved. Accordingly, at time $k = 0$, to find the adjusted solution $x^a(1/0)$, the control $x^e(1/0)$ is first determined by omitting the non-negativity constraints. From (5.21), (5.22), $\phi_{1/0}$ given by (5.25a) and $\Gamma_{1/0}$ given by (5.26a), it may be checked that the first single-stage control setting is

$$x^e(1/0) = \begin{bmatrix} 1.1784 \\ 3.8216 \end{bmatrix}$$

and, as it also satisfies the non-negativity control restrictions, it immediately follows that $x^e(1/0) = x^a(1/0)$.

Dual simplex method Supposing that the random term in (5.23) for

Tableau 5.1 Initial tableau.

	z	β_1	β_2
	0	0	0
x_1	−5.221	(−0.4067)	0.4067
x_2	13.221	0.4067	−0.4067
β_1	0	−1	0
β_2	0	0	−1

period one is $u_1 = 0$, and $x^a(1/0)$ is as specified, then $Y_1' = (4.8870, 3.4766)$. Before $x^a(2/1)$, the optimal control solution to the second single-stage control problem, is computed the parameter estimates given at time $k = 0$ are revised in the manner dictated by the equations in Section 5.2. Then, omitting the non-negativity constraints, and from (5.21) and (5.22), it turns out that the second stage control setting is

$$x^e(2/1) = \begin{bmatrix} -5.2210 \\ 13.2210 \end{bmatrix}$$

As $x^e(2/1)$ does not satisfy the non-negativity constraints, it must be adjusted for feasibility. Using the dual simplex algorithm to yield $x^a(2/1)$, and performing the necessary computations, the initial tableau is given by Tableau 5.1 for the dual simplex method. Note that $x^e(2/1)$ furnishes the initial primal infeasible solution. Applying the usual rules of choosing a variable to leave the basis and a variable to enter, the pivot element is obtained and is shown circled. This yields the next tableau, Tableau 5.2, which contains the solution to the problem with non-negativity constraints, i.e.

$$x^a(2/1) = \begin{bmatrix} 0.0000 \\ 8.0000 \end{bmatrix}$$

Tableau 5.2 Final tableau.

	z	x_1	β_2
	0	0	0
x_1	0		
x_2	8.0000		
β_1	12.8375		
β_2	0		

Method of imposing constraints If the alternative method of imposing constraints is employed, then the control solutions are found to be

$$x^{ic}(1/0) = \begin{bmatrix} 0 \\ 5 \end{bmatrix} \quad \text{and} \quad x^{ic}(2/0) = \begin{bmatrix} 0 \\ 8 \end{bmatrix}$$

To see how they are obtained we note from the above that

$$x^{e}(1/0) = \begin{bmatrix} -0.1110 \\ 5.1110 \end{bmatrix} \quad \text{and} \quad x^{e}(2/0) = \begin{bmatrix} -2.3024 \\ 10.3024 \end{bmatrix}$$

Hence, accordingly we impose the constraints $x_{1,2} = 0$, where $x_{1,2}$ is the first component of x at time $t = 2$. This means that at time 2 the constraint set $F_2 x_2 = f_2$ is

$$\begin{bmatrix} 1 & 1 \\ 1 & 0 \end{bmatrix} x_2 = \begin{bmatrix} 8 \\ 0 \end{bmatrix}$$

while $F_1 x_1 = f_1$ remains unchanged. On solving this new problem we get

$$x^{e}(1/0) = \begin{bmatrix} -0.7143 \\ 5.7143 \end{bmatrix} \quad \text{and} \quad x^{e}(2/0) = \begin{bmatrix} 0 \\ 8 \end{bmatrix}$$

For the next iteration, we need to impose the constraint $x_{1,1} = 0$. Clearly the resulting control solution must be

$$x^{ic}(1/0) = \begin{bmatrix} 0 \\ 5 \end{bmatrix} \quad \text{and} \quad x^{ic}(2/0) = \begin{bmatrix} 0 \\ 8 \end{bmatrix}$$

It turns out that, if all unknown variables are replaced by their expected values, then

$$\begin{pmatrix} \text{loss function when} \\ x^{a} \text{ used} \end{pmatrix} - \begin{pmatrix} \text{loss function when} \\ x^{ic} \text{ used} \end{pmatrix}$$
$$= 97.9642 - 48.7160 = 49.2482$$

This implies that the method of imposing constraints gives an improved control performance of the system. In both cases note the difference between the solutions with and without the non-negativity constraints.

5.6 *Conclusion*

This chapter has considered the problem of controlling a linear dynamic system where the parameters are not only unknown but are also time-varying, and subject to constraints on the control variables. The case where only linear equality constraints are present is first solved. By using the approach of Farison *et al.* (1967), separation of the estimation and control portions of the problem is achieved. The necessary stages involved in arriving at the solutions may be summarised as follows:

(a) At each time period k, determine estimates of the mean vector and covariance matrices of the unknown parameters for the future, conditional on the present system model. These future estimates are given by (5.10) and (5.11).
(b) Determine the quantities $\hat{G}_{t/k}$ and $\hat{g}_{t/k}$ for $t = T, \ldots, k + 1$ according to (5.17) and (5.18).
(c) From (5.16) the optimal control setting $x^e(k + 1/k)$ at time k may be evaluated.
(d) When $x^e(k + 1/k)$ is used and Y_{k+1} becomes available, the system model is updated. This is accomplished via Equations (5.8d) and (5.8e).
(e) With the index k increased to $k + 1$, steps (a)–(d) are repeated, yielding $x^e(k + 2/k + 1)$, the optimal control setting at time $k + 1$. Hence, by repeated minimisations, the method yields the sequence of control inputs $x^e(1/0), \ldots, x^e(T/T - 1)$.

The addition of non-negativity control constraints to the problem has also been discussed in this chapter. A method proposed for solution in Section 5.4 allows a solution to be found by considering a sequence of single-stage optimising problems. These solutions are 'single-stage optimisers' which are also feasible solutions to the multi-stage problem. The steps to obtain the single-stage optimal solutions are

(a) At time k ($k = 0, \ldots T - 1$), based on the current system model, determine $x^e(k + 1/k)$, the solution to the single-stage problem when only linear equality constraints are imposed. This solution $x^e(k + 1/k)$ is found as described earlier.
(b) When $x^e(k + 1/k)$ does not satisfy the non-negativity constraints, then it must be adjusted. As before, the dual simplex algorithm provides the necessary adjustment mechanism to yield the optimal solution $x^a(k + 1/k)$. An alternative method suggested is that of imposing constraints. This adds a control constraint at the zero level at each iteration until no control variable is negative.

To illustrate the various computations needed to arrive at the solutions, a simple numerical example demonstrates that the control solutions are quite different from the equality- and inequality-constrained cases. Furthermore, there is an improvement in the control performance when updating of parameter estimates is allowed.

In the next chapter we shall consider not only control problems with a general linear inequality control constraint but also constraints on the state variables. Again, it will be seen that the solution strategy used above is still applicable; the problem is first solved with only linear equality constraints present, then it is adjusted for feasibility to satisfy the other inequality constraints.

6 *Spatial control with constraints on both state and control variables*

6.1 State and control constraints

In earlier chapters optimal control problems have been considered with constraints on only the control variables. In this chapter the optimal control problem is extended to situations in which there are equality and inequality constraints on the state variables as well. The proposed two-part solution strategy turns out to be applicable with modifications to the initial tableau for the dual simplex method. The solution techniques described in earlier chapters therefore provides a general approach to solving optimal control problems with constraints on both state and control variables.

The addition of inequality constraints to the state variables in the constrained optimal control problem provides a framework for solving a number of relevant practical problems. For example, in geographical applications of national settlement system planning (MacKinnon 1975, Propoi & Willekens 1978), the state variables (which denote the population in each region) are modified by inflows of migrants. The control problem is then to allow only those levels of migration which prevent the level of population in each region from falling below some critical target level. Thus there are inequality constraints on the state variables of population. Such constraints could also be meaningful within the context of grant distribution between the two levels of government in a nation. The state variables representing some aggregate measure of the prevailing economic conditions, for example, the variables total expenditure and total revenue, might be constrained to allow for expenditure exceeding revenue in some areas.

Inequality state constraints can also be used to overcome criticisms in the use of the quadratic objective function. It is acknowledged that this objective function does not discriminate between positive or negative deviations from the desired target setting of the state variables. In macro-economic control, for example, if the unemployment rate is set at a desired level of 6%, an actual outcome of 5% is given the same weighting in the objective function as an outcome of 7%. Imposition of a state constraint, that the unemployment rate must not be more than 6%, restricts the objective function by not allowing positive deviations to be considered. In other words, attention can be restricted to cases of deviations from targets in only one direction.

When the set of constraints is consistent and an optimal solution exists, the control solution described in this chapter will always be the true optimum. For this reason, the method is an improvement over existing solution techniques which aim to find only a local optimum (Ritch 1973), what is frequently termed an 'engineering solution' (Owens & Jones 1978). The emphasis of these latter techniques is on obtaining only a feasible solution which of course may sometimes give the true optimal solution, but there is no guarantee that this will be the case.

Section 6.2 considers the control problem with the usual linear equality constraints on the control variables as well as the more general inequality constraint $R_t x_t \geqslant r_t$, where x_t is the control vector at time t. It turns out that the discussion in this section also provides the solution to the control problem with state inequality constraints considered in Section 6.3. In this chapter only deterministic models are considered.

6.2 *Control solutions for general inequality constraints*

This section generalises the analysis of Section 3.3 from non-negativity control constraints to the more general inequality constraints

$$R_t x_t \geqslant r_t \qquad \text{for } t = 1, \ldots, T \tag{6.1}$$

where R_t is a known coefficient matrix. Of course, when $R_t = I$, the identity matrix, and $r_t = 0$ for each time period, then (6.1) reduces to the non-negativity constraints (3.30b).

The constraint (6.1), with the inequality sign reversed, may be interpreted as a 'budget' constraint. For instance, if certain costs are incurred when the controls are utilised, then the (i, j)th element of R_t denotes this cost. The quantity r_t represents the budget available at time t for imposing the control setting. Such budgetary constraints are encountered in the national settlement system planning problem (Propoi & Willekens 1978) and it may be seen that the form (6.1) is a generalisation of the budget constraints considered by them.

With $r_t = 0$ and a suitable definition of R_t, the constraint (6.1) allows for instances in which not all the control components are to be non-negative. The constraint (3.30b) does not give this flexibility. For example, if the dimension of x_t is four and it is only meaningful to restrict the first two components to be non-negative, then this could be modelled by choosing

$$R_t = \begin{bmatrix} 1 & 0 & 0 & 0 \\ 0 & 1 & 0 & 0 \end{bmatrix}$$

The optimal control problem that must be solved may be formally stated as follows:

Given the linear dynamic model (2.13) *with the initial condition* Y_0, *the optimal control problem is to find the sequence of control variables* $x_1, \ldots, x_T$ *that minimises the quadratic objective function* (2.15) *when the control variables are subject to the following set of constraints for each time period t:*

$$F_t x_t = f_t \tag{6.2a}$$

$$R_t x_t \geq r_t \tag{6.2b}$$

In (6.2), F_t and R_t are known matrices of dimension $k \times m$ and $l \times m$ respectively. It is assumed that k is less than m but l is not restricted. The vectors f_t and r_t are known specified quantities.

Compared to Section 3.3, no additional difficulties are encountered in obtaining the optimal control solution to the above problem. The usual method of dynamic programming followed by an adjustment mechanism based on the dual simplex method is still applicable. The necessary modifications are now given.

The first step of this two-part solution strategy considers the control problem with only the linear equality constraints (6.2a) present. Clearly, this can be solved by using the method of dynamic programming as described in Section 3.2. When the resulting solutions $x_1^{\circ}, \ldots, x_T^{\circ}$ do not satisfy the constraints (6.2b) then the second part of the solution strategy becomes operative. This simultaneously adjusts $x_1^{\circ}, \ldots, x_T^{\circ}$ for feasibility as a linear programming problem. Following Section 3.3 and defining

$$J_4(y) = \begin{bmatrix} R_1 x_1 - r_1 \\ \vdots \\ R_T x_T - r_T \end{bmatrix}$$

the static programming problem considers minimising the quadratic objective function (2.15) subject to the constraints $J_1(y) = 0$, $J_2(y) = 0$ and $J_4(y) \geq 0$, where y, $J_1(y)$ and $J_2(y)$ are given previously in Section 3.3. Using the method of Lagrange multipliers, first form the Lagrangian

$$L = \tfrac{1}{2} W + \alpha' J_1(y) + \lambda' J_2(y) - \gamma' J_4(y) \tag{6.3}$$

where α and λ are both unrestricted multipliers, but γ is non-negative. With $X = (x'_1, \ldots, x'_T)'$, $R = \text{diag}(R_1, \ldots, R_T)$, which is a block diagonal matrix, we may go through similar calculations as in Section 3.3. After these calculations it may be shown that the optimal values of X and γ are related according to

$$X = \hat{G}_s Y_0 + \hat{g}_s + \bar{B}R'\gamma \tag{6.4}$$

The quantities $\hat{G}_s$, $\hat{g}_s$ and $\bar{B}$ are as defined in Section 3.3. Furthermore, the optimal controls X^* and optimal Lagrange multipliers γ^* must also satisfy the following conditions from Kuhn–Tucker theory for non-linear programming:

$$J_4^*(y) \geqslant 0, \ \ \gamma^* \geqslant 0 \ \ \text{and} \ \ \gamma^{*\prime} J_4(y) = 0 \tag{6.5}$$

There is no need to compute $\hat{G}_s$ and $\hat{g}_s$ in (6.4) because it turns out that $\hat{G}_s Y_0 + \hat{g}_s = \bar{G}_s$, which is identical to $(x_1^{\circ\prime}, \ldots, x_T^{\circ\prime})'$, the optimal control solutions obtained from the first step of the strategy. When the inequality constraint (6.2b) is violated, it will be necessary to adjust the first step of the solution for feasibility. This may be achieved by considering the following programming problem:

$$\text{maximize } z = 1'_v(FX - f) \tag{6.6}$$

subject to

$$\begin{aligned} \theta - R\bar{B}R'\gamma &= R\bar{G}_s - r \\ \gamma, \theta &\geqslant 0 \\ \gamma'\theta &= 0 \end{aligned}$$

where $r = (r'_1, \ldots, r'_T)'$ and θ is a vector of 'surplus' variables that converts the constraint $J_4(y) \geqslant 0$ into the equality form. Thus the complementary slackness condition $\gamma' J_4(y) = 0$ is equivalent to $\gamma'\theta = 0$. Except for this restriction the problem is linear. As $\bar{G}_s$ is already available, the initial basic solution to (6.6) is obtained by setting $\gamma = 0$ to give $\theta = R\bar{G}_s - r$. It is assumed that this initial solution is infeasible, otherwise the control problem with constraints (6.2) is solved. However, although this solution is primal infeasible, it is dual feasible. Hence, the dual simplex method may again be used to determine the true optimal solution although a variation to this algorithm is needed to handle the complementary slackness condition. This variation is similar to that described in Section 3.3 and is based on Wolfe's (1959) algorithm for solving quadratic programming problems. When an optimal solution exists, this strategy will yield the true optimal solution in a finite number of steps. Note that the set of equality constraints appearing in (6.6) is obtained from (6.4) in the following way:

$$RX = R\bar{G}_s + R\bar{B}R'\gamma$$

but $RX - r - \theta = 0$ so that

$$\theta - R\bar{B}R'\gamma = R\bar{G}_s - r. \tag{6.7}$$

6.3 *State inequality constraints*

In some practical applications of optimal control techniques, it may be necessary to impose inequality constraints on the state variables, e.g. in national settlement system planning (see Ch. 8 and Propoi & Willekens 1978) and in the allocation of intergovernmental grants (see Ch. 7). In order that the subsequent exposition is more general, consider the following optimal control problem with constraints on both state and control variables:

Given the linear dynamic model (2.13) *with the initial condition* Y_0, *the optimal control problem is to find the set of control variables* $x_1, \ldots, x_T$ *that minimises the quadratic objective function* (2.15) *when the control and state variables are subject to the following constraints for each time period t:*

$$F_t x_t = f_t \tag{6.8a}$$

$$R_t x_t \geqslant r_t \tag{6.8b}$$

$$S_t Y_t \geqslant s_t \tag{6.8c}$$

The constraints (6.8a) and (6.8b) are the same as (6.2a) and (6.2b) respectively. In (6.8c), S_t is a known matrix of dimension $k_2 \times p'$, where $p' = pn + qm$ and s_t is a $k_2 \times 1$ vector which is known. There is no restriction imposed on k_2. Note that, when $S_t = I$ and $s_t = a_t$ for each time period t, then (6.8c) ensures that, in the quadratic objective function (2.15), only positive deviations are permitted, thereby avoiding one of the criticisms of the quadratic loss function.

Instead of incorporating the states inequality constraint (6.8c) directly into the solution strategy, it is more convenient to transform (6.8c) into inequality constraints on the control variables. Thus, the control problem will then have only constraints on the controllers and will allow immediate application of the results obtained in Section 6.2.

Now, for each time period $t = 1, \ldots, T$ the constraints on the state variables are given by

$$\begin{aligned} S_1 Y_1 &\geqslant s_1 \\ S_2 Y_2 &\geqslant s_2 \\ &\vdots \\ S_T Y_T &\geqslant s_T \end{aligned} \tag{6.9}$$

Using the linear dynamic model (2.13) the above constraint for the first time period becomes

$$S_1 A_1 Y_0 + S_1 B_1 x_1 + S_1 b_1 \geqslant s_1$$

or, equivalently,

$$S_1 B_1 x_1 \geqslant s_1 - S_1 A_1 Y_0 - S_1 b_1 \tag{6.10}$$

Similarly, the state constraint for the second time period may be expressed as

$$S_2 A_2 Y_1 + S_2 B_2 x_2 + S_2 b_2 \geqslant s_2$$

and using the system model (2.13) once again we get

$$S_2 A_2 A_1 Y_0 + S_2 A_2 B_1 x_1 + S_2 B_2 x_2 + S_2 b_2 + S_2 A_2 b_1 \geqslant s_2$$

or

$$S_2 A_2 B_1 x_1 + S_2 B_2 x_2 \geqslant s_2 - S_2 A_2 A_1 Y_0 - S_2 b_2 - S_2 A_2 b_1 \tag{6.11}$$

It is now clear that the set of state constraints (6.9) is equivalent to the following constraints imposed only on the control variables:

$$\begin{bmatrix} S_1 B_1 & & & & \\ S_2 A_2 B_1 & S_2 B_2 & & & 0 \\ \vdots & & & & \\ S_T A_T \ldots A_2 B_1 & S_T A_T \ldots A_3 B_2 & \ldots & & S_T B_T \end{bmatrix} \begin{bmatrix} x_1 \\ x_2 \\ \vdots \\ x_T \end{bmatrix}$$

$$\geqslant \begin{bmatrix} s_1 - S_1 A_1 Y_0 \\ s_2 - S_2 A_2 A_1 Y_0 \\ \vdots \\ s_T - S_T A_T \ldots A_1 Y_0 \end{bmatrix} - \begin{bmatrix} S_1 & & 0 \\ & S_2 & \\ & & \ddots \\ 0 & & & S_T \end{bmatrix}$$

$$\times \begin{bmatrix} b_1 \\ A_2 b_1 + b_2 \\ \vdots \\ A_T \; \ldots \; A_2 b_1 + A_T \; \ldots \; A_3 b_2 + \ldots + b_T \end{bmatrix} \tag{6.12a}$$

or, more conveniently,

$$\bar{S}X \geqslant \bar{s} \tag{6.12b}$$

with obvious definitions for $\bar{S}$ and $\bar{s}$. Note that the expression on the right hand side of the inequality (6.12a) consists of known quantities. Hence, the term $\bar{s}$ is known.

Again, a two-step strategy can be used to determine the optimal solution. As previously, the first step solves the optimal control problem when only the equality constraint (6.8) is present. If this first step solution also satisfies the inequality constraints (6.8b) and (6.8c), then we have found the optimal solution required and the second step of the solution strategy is not necessary. Otherwise the second step is needed and its purpose is to adjust the first step solution to satisfy the other constraints. The adjustment is again based on the dual simplex algorithm and the details are now given.

With the definitions for $y, J_1(y), J_2(y), J_4(y)$ given previously, we now define

$$J_5(y) = \begin{bmatrix} \bar{S}_{11}x_1 - \bar{s}_1 \\ \vdots \\ \bar{S}_{T1}x_1 + \ldots + \bar{S}_{TT}x_T - \bar{s}_T \end{bmatrix}$$

where $\bar{S}_{ij}$ is the (i, j)th element of $\bar{S}$ and $\bar{s}_t$ is the tth component of $\bar{s}$. The static programming problem now calls for the objective function (2.15) to be minimised subject to $J_1(y) = 0$, $J_2(y) = 0$, $J_4(y) \geqslant 0$, $J_5(y) \geqslant 0$. This may be done by forming the Lagrangian

$$L = \tfrac{1}{2}W + \alpha' J_1(y) + \lambda' J_2(y) - \gamma' J_4(y) - \nu' J_5(y) \tag{6.13}$$

where the multipliers γ and ν are restricted to be non-negative. It follows from the Kuhn–Tucker conditions for an optimum, after some simplifications, that

$$X = \bar{G}_s + \bar{B}R'\gamma + \bar{B}\bar{S}'\nu \tag{6.14}$$

where $\bar{G}_s$, $\bar{B}$, R and X are previously defined. In addition, the following conditions must also hold for an optimum:

$$\begin{aligned} J_4(y) \geq 0, \quad \gamma \geq 0 \quad \text{and} \quad \gamma' J_4(y) = 0 \\ J_5(y) \geq 0, \quad \nu \geq 0 \quad \text{and} \quad \nu' J_5(y) = 0 \end{aligned} \tag{6.15}$$

The vector $\bar{G}_s$ is the solution obtained from the first part of the strategy. To adjust $\bar{G}_s$ for feasibility, we need to solve the following programming problem:

$$\text{maximise } z = 1'_\nu (FX - f)$$

subject to

$$\begin{aligned} \theta - R\bar{B}R'\gamma - R\bar{B}\bar{S}'\nu &= R\bar{G}_s - r \\ \Omega - \bar{S}\bar{B}R'\gamma - \bar{S}\bar{B}\bar{S}'\nu &= \bar{S}\bar{G}_s - \bar{s} \\ \gamma, \nu, \theta, \Omega &\geq 0 \\ \gamma'\theta &= 0 \\ \nu'\Omega &= 0 \end{aligned} \tag{6.16}$$

where Ω is the vector of surplus variables which changes $J_5(y) \geq 0$ into an equality. Setting $\gamma = 0$ and $\nu = 0$ gives the initial basic solution as $\theta = R\bar{G}_s - r$ and $\Omega = \bar{S}\bar{G}_s - \bar{s}$, which by assumption is infeasible. Fortunately, it is dual feasible and hence the programming problem (6.16) can be solved using the dual simplex method with the usual variation that is needed to allow for the presence of the complementary slackness conditions $\gamma'\theta = 0$ and $\nu'\Omega = 0$. If the set of constraints (6.8) is consistent and a solution exists, the optimal control solution will be determined in a finite number of steps.

In passing, an interesting special case arises from the above analysis relating to the formulation of the national settlement system planning problem given by, e.g. Mehra (1975), MacKinnon (1975) and Propoi and Willekens (1978), and discussed in Chapter 8. Instead of imposing a budget constraint for each time period, for which the constraint (6.2b) is a generalisation, a global budgetary constraint is sometimes used. For generality, a budgetary constraint for the whole planning period may be expressed by

$$Q_1 x_1 + \ldots + Q_T x_T \geq \bar{q} \tag{6.17}$$

This constraint is a special case of (6.12a) where $\bar{S} = [Q_1, \ldots, Q_T]$ and $\bar{s} = \bar{q}$. Hence there is no difficulty in deriving the optimal solution with equations such as (6.17) replacing (6.12b).

6.4 *An example*

To illustrate the calculations involved in determining the optimal control solution by the two-stage strategy discussed above, the example of a two-region economy of Section 3.3 is extended to include state inequality constraints as well as control (budgetary) constraints. Specifically, the optimal control problem is

$$\text{minimise} \sum_{t=1}^{2} (Y_t - a_t)'M_t(Y_{\cdot t} - a_t)$$

subject to

$$Y_t = \begin{bmatrix} 2 & 1 \\ 0 & 2 \end{bmatrix} Y_{t-1} + \begin{bmatrix} 1 & 0 \\ 2 & 1 \end{bmatrix} x_t$$

$$F_t x_t = f_t$$

$$R_t x_t \geqslant r_t, \qquad t = 1, 2$$

$$Y_t \geqslant a_t$$

where $Y_0 = (3, -1)'$, the targets are defined by $a_1 = (5, 3)'$, $a_2 = (8, 6)'$, the quadratic cost weighting is given by $M_1 = M_2 = I$, the constraint matrices are $F_1 = F_2 = (1, 1)$,

$$R_1 = [-1 \quad -2], \qquad R_2 = [-1 \quad -3],$$

with constraint limits $f_1 = 5$, $f_2 = 8$, $r_1 = -8$, $r_2 = -12$. Note that, for simplicity, the non-negativity constraints on the control variables are not imposed in this example.

The first part of the solution strategy solves the control problem with only the equality constraint imposed. For the computations in Section 3.3 the resulting control solution is

$$x_1^\circ = [0.6, 4.4], \qquad x_2^\circ = [-8, 16].$$

Clearly, x_1° and x_2° do not satisfy the inequality constraints. Hence, the second step of the solution strategy is required. First, note that the state inequality constraints

$$Y_1 \geqslant a_1, \qquad Y_2 \geqslant a_2$$

become, after the necessary transformations,

$$\begin{bmatrix} B_1 & 0 \\ A_2 B_1 & B_2 \end{bmatrix} \begin{bmatrix} x_1 \\ x_2 \end{bmatrix} \geqslant \begin{bmatrix} a_1 - A_1 Y_0 \\ a_2 - A_2 A_1 Y_0 \end{bmatrix}$$

It is already known (from Section 3.3) that

$$\bar{B} = \frac{1}{5}\begin{bmatrix} 2 & -2 & -5 & 5 \\ -2 & 2 & 5 & -5 \\ -5 & 5 & 15 & -15 \\ 5 & -5 & -15 & 15 \end{bmatrix}$$

Now,

$$R = \begin{bmatrix} -1 & -2 & 0 & 0 \\ 0 & 0 & -1 & -3 \end{bmatrix}, \qquad r = \begin{bmatrix} -8 \\ -12 \end{bmatrix}$$

and

$$\bar{S} = \begin{bmatrix} 1 & 0 & 0 & 0 \\ 2 & 1 & 0 & 0 \\ 4 & 1 & 1 & 0 \\ 4 & 2 & 2 & 1 \end{bmatrix}, \qquad \bar{s} = \left[\begin{array}{c} 0 \\ 5 \\ \hline 0 \\ 10 \end{array}\right]$$

Therefore,

$$R\bar{G}_s - r = \begin{bmatrix} -9.4 \\ -40 \end{bmatrix} - \begin{bmatrix} -8 \\ -12 \end{bmatrix} = \begin{bmatrix} -1.4 \\ -28 \end{bmatrix}$$

$$\bar{S}\bar{G}_s - \bar{s} = \left[\begin{array}{c} 0.6 \\ 5.6 \\ \hline -1.2 \\ 11.2 \end{array}\right] - \left[\begin{array}{c} 0 \\ 5 \\ \hline 0 \\ 10 \end{array}\right] = \left[\begin{array}{c} 0.6 \\ 0.6 \\ \hline -1.2 \\ 1.2 \end{array}\right]$$

Furthermore,

$$\bar{B}R' = \frac{1}{5}\begin{bmatrix} 2 & -10 \\ -2 & 10 \\ -5 & 30 \\ 5 & -30 \end{bmatrix}, \qquad \bar{B}\bar{S}' = \frac{1}{5}\begin{bmatrix} 2 & 2 & 1 & -1 \\ -2 & -2 & -2 & 1 \\ -5 & -5 & 0 & 5 \\ 5 & 5 & 0 & -5 \end{bmatrix}$$

$$R\bar{B}R' = \frac{1}{5}\begin{bmatrix} 2 & -10 \\ -10 & 60 \end{bmatrix}, \qquad \bar{S}\bar{B}\bar{S}' = \frac{1}{5}\begin{bmatrix} 2 & 2 & 1 & -1 \\ 2 & 2 & 1 & -1 \\ 1 & 1 & 3 & 2 \\ -1 & -1 & 2 & 3 \end{bmatrix}$$

$$R\bar{B}\bar{S}' = \frac{1}{5}\begin{bmatrix} 2 & 2 & 1 & -1 \\ -10 & -10 & 0 & 10 \end{bmatrix}.$$

The initial tableau for the dual simplex algorithm can now be easily obtained. This is given in Tableau 6.1. It may be seen that θ_2 is chosen to leave the basis and its place taken by γ_2; the pivot element is shown circled. This yields Tableau 6.2. Now the variable to leave the basis is θ_1 and the variable to enter is γ_1, with the pivot element circled: Tableau 6.3 follows.

Tableau 6.1 Initial tableau.

	z	γ_1	γ_2	ν_1	ν_2	ν_3	ν_4
	0	0	0	0	0	0	0
θ_1	−1.4	−0.4	2.0	−0.4	−0.4	−0.2	0.2
θ_2	−28	2.0	−12.0	2.0	2.0	0.0	−2.0
Ω_1	0.6	−0.4	2.0	−0.4	−0.4	−0.2	0.2
Ω_2	0.6	−0.4	2.0	−0.4	−0.4	−0.2	0.2
Ω_3	−1.2	−0.2	0.0	−0.2	−0.2	−0.6	−0.4
Ω_4	1.2	0.2	−2.0	0.2	0.2	−0.4	−0.6
γ_1	0	−1	0	0	0	0	0
γ_2	0	0	−1	0	0	0	0
ν_1	0	0	0	−1	0	0	0
ν_2	0	0	0	0	−1	0	0
ν_3	0	0	0	0	0	−1	0
ν_4	0	0	0	0	0	0	−1

Tableau 6.2 After first iteration.

	z	γ_1	θ_2	ν_1	ν_2	ν_3	ν_4
	0	0	0	0	0	0	0
θ_1	−91/15	−1/15	1/6	−1/15	−1/15	−1/5	−2/15
θ_2	0	0	−1	0	0	0	0
Ω_1	−61/15	−1/15	1/6	−1/15	−1/15	−1/5	−2/15
Ω_2	−61/15	−1/15	1/6	−1/15	−1/15	−1/5	−2/15
Ω_3	−6/5	−1/5	0	−1/5	−1/5	−3/5	−2/5
Ω_4	88/15	−2/15	−1/6	−2/15	−2/15	−2/5	−4/15
γ_1	0	−1	0	0	0	0	0
γ_2	7/3	−1/6	−1/12	−1/6	−1/6	0	1/6
ν_1	0	0	0	−1	0	0	0
ν_2	0	0	0	0	−1	0	0
ν_3	0	0	0	0	0	−1	0
ν_4	0	0	0	0	0	0	−1

Tableau 6.3 Final tableau showing optimal solution.

	z	θ_1	θ_2	ν_1	ν_2	ν_3	ν_4
	0	0	0	0	0	0	0
θ_1	0						
θ_2	0						
Ω_1	2						
Ω_2	2						
Ω_3	17						
Ω_4	18						
γ_1	91						
γ_2	17.5						
ν_1	0						
ν_2	0						
ν_3	0						
ν_4	0						

As the first column of Tableau 6.3 does not have any negative entries, we have arrived at the optimal solution. Hence, there is no need to calculate the other entries in the tableau. However, note that the optimal solution is not given in terms of the control variables X; it must be calculated according to (6.14), yielding the optimal X after substituting the optimal values of γ and ν. This gives the optimal controls X^* as

$$X^* = \begin{bmatrix} 0.6 \\ 4.4 \\ -8 \\ 16 \end{bmatrix} + \begin{bmatrix} 1.4 \\ -1.4 \\ 14 \\ -14 \end{bmatrix} + \begin{bmatrix} 0 \\ 0 \\ 0 \\ 0 \end{bmatrix} = \begin{bmatrix} 2.0 \\ 3.0 \\ 6.0 \\ 2.0 \end{bmatrix}$$

or

$$x_1^* = (2, 3) \quad \text{and} \quad x_2^* = (6, 2)$$

Although the non-negativity constraints on the control variables were not imposed, the optimal controls turn out to be non-negative. Thus, the above solution will also be the optimal control setting for the problem with the constraints $x_1 \geqslant 0$ and $x_2 \geqslant 0$ added.

6.5 *Conclusion*

For deterministic linear dynamic models in the presence of inequality constraints on both the state and control variables the true optimal sequence of control settings that minimise the usual quadratic objective function can be found. In this chapter, it has been demonstrated how the two-step solution strategy of Chapter 3 is still applicable. The only change necessary is in deriving the initial tableau used to adjust the first step solution for feasibility in the dual simplex method. To summarise, the steps involved are as follows:

Step 1 Solve the optimal control problem with only the linear equality constraints on the control variables present by the recursive algorithm described in Section 3.2. Denote this solution by $\overset{\circ}{x}_1, \ldots, \overset{\circ}{x}_t$.

Step 2 Check if $x_1^\circ, \ldots, x_T^\circ$ satisfy the inequality constraints of the original problem. If the answer is yes, then $x_1^\circ, \ldots, x_T^\circ$ is the optimal solution required. Otherwise go to step 3.

Step 3 Calculate the necessary matrices and vectors and form the initial tableau for solving the static linear programming problem. Obtain the initial basic primal infeasible solution by setting $\nu = 0$ and $\gamma = 0$. Then use the dual simplex algorithm to proceed to the optimal solution. It may be necessary to use (6.14) or (6.4), depending on whether any state inequality constraint is imposed, to obtain the required sequence of optimal control settings $x_1^*, \ldots, x_T^*$.

In using the dual simplex method, the usual variation is introduced to ensure that the complementary slackness condition holds.

The imposition of inequality state and control constraints provides a framework in which interesting and important spatial control problems may be solved. It has been shown in this chapter that solving a control problem with inequality state constraints is equivalent to solving one with just constraints on the control variables. With appropriate specification of the state inequality constraints, it is possible to avoid the major criticism of the quadratic objective function, namely its equal treatment of positive and negative deviations. The two-step solution strategy (first encountered in Ch. 3) seems to provide a unifying method to solve linear-quadratic optimal control problems with constraints on both the state and control variables.

Consideration of constrained control problems in this and the previous chapters has been analytical. The applicability and relevance of these theoretical developments will be discussed in the following chapters, which are concerned with developing a number of large-scale applications.

PART II APPLICATIONS

7 *Intergovernmental grant allocation*

7.1 Intergovernmental grants

One of the major applications of control theory to spatial systems which has been developed to date has been for the allocation of intergovernmental grants. All Western countries possess systems of local government in addition to a central or federal government. In some countries, especially those which have federal constitutions, the local and central levels of government are supplemented by intermediate levels of government: the states in the USA and Australia, the provinces in Canada, the Länder in Austria and Germany, and the cantons in Switzerland. Within such multilevel structures there is usually a complex mechanism of intergovernmental flows of financial resources. Sometimes this is from local level to higher level governments; sometimes the flow is between governments at the same level; most commonly the flow of resources is either from central/federal governments to state and local governments or from states to local governments.

This intergovernmental grant allocation problem has been extensively treated in economics (see Oates 1972, Musgrave & Musgrave 1980) as well as geography (see Bennett 1980, 1982a). This chapter develops solutions to the grant allocation problem using the methods of optimum control theory and then presents empirical results of applying this theory to allocation of the Rate Support Grant (RSG) in Britain.

The criteria which can be used for allocating intergovernmental grants concern four main variables: the local tax base, the local tax rate, the level of local financial need, and the local expenditure level. Based on these variables, the allocation of grants using the methods of optimum control theory seeks to encompass five main criteria into grant allocations, namely

(a) *Equalisation*. This concerns the attempt to place local governments in an equal position to provide a given service level at a given cost. Two criteria of equalisation are employed in the RSG, separately up to 1980/1, and simultaneously thereafter. First, local governments are compensated for differences in levels of expenditure need. These arise because of different sizes of client group, costs of servicing different groups, or differences in the economics of service delivery. A second criterion is resource equalisation, which aims to compensate areas for differences in the size of their tax base. Tax base must be assessed against the revenue sources available and hence may bear little relation to the distribution of personal wealth. In Britain the tax base is the local rateable value available for taxation by the rates (a property tax).

(b) *Stability*. This requires that grants maintain a relatively stable pattern of distribution from one time period to another and that changes in distribution result only from objective changes in circumstances. This is important in local expenditure planning and to prevent large fluctuations in local tax rates.

(c) *Certainty*. This requires that once grants are allocated the total sum remains unchanged. In the allocation of the RSG, in particular, considerable uncertainty has been introduced through inconsistent use of Increase Orders after initial allocations have been made. This is also important in local expenditure planning.

(d) *Dynamics*. This is a very important criterion and requires that past and future local decisions are considered in present grant allocation. Past expenditure patterns are important in affecting future needs, and future needs should affect present allocations. Perhaps the best example is in the analysis of population cohorts: knowledge of the present age distribution of a population gives good potential to forecast, e.g. the future needs for schools of different categories, or the need for services for the aged. Small present needs may therefore sometimes justify a large grant allocation in order to cater for large future expenditures, and vice versa.

(e) *Vertical (hierarchical) criteria*. Vertical allocation criteria concern fiscal balance: the redistribution of revenue between levels of government to meet hierarchical differences in expenditure requirements. Many public services can be efficiently organised only at local level, but most of the high yielding taxes are usually best reserved to central governments. Hence, resources must be redistributed between levels of government to achieve fiscal balance.

Other criteria are also often important in grant allocation, e.g. their use for compensation of benefit spillover (Pigou 1947), their use to improve social distribution (Musgrave & Musgrave 1980), or their use to improve local accountability and efficiency in service delivery. However, the five criteria listed above are the most important from the point of view of control theory.

7.2 *The Rate Support Grant (RSG) in England and Wales*

The Rate Support Grant (RSG) is by far the largest intergovernmental grant which is used in the United Kingdom and accounts for 80% of all such grants. It is a general or 'block' grant which can be used relatively flexibly by local authorities. However, it is distributed taking account of the expenditure requirements and local tax capacity of different local governments. Initiated in 1967/8 various formulae have been applied to its allocation, many terms of which date back to 1929 (see Committee of Enquiry into Local Government Finance 1976, Bennett 1982a). As a result, the grant structure has become an exceedingly complex programme which has evolved over a long period to encompass many objectives which differ from those explicitly stated at its inception. Similarly, like most public policy programmes, it has encompassed an 'inherited' pattern of allocation deriving from previous methods of allocating grants to local governments (see, for example, Boyle 1966, Committee of Enquiry into Local Government Finance 1976, Foster *et al.* 1980, Bennett 1982a).

Over the period from 1967/8 to 1980/1 the grant was divided into three separate elements: needs, resources and domestic elements. The *needs element* accounted for about 60% of the total and was aimed at allocating to local authorities an amount equal to their expenditure requirements. It was assessed according to a number of needs criteria which were based on three sets of factors: first, variation in the proportion of people who consumed each local authority's services (e.g. the number of old people, the number of people in various levels of education, etc.); secondly, variation in the quantity of inputs required to provide services in each local authority (e.g. variations in pupil–teacher ratios between urban and rural areas); thirdly, variations in costs of inputs in different local authorities (e.g. salary weightings in London and the south-east of England). From 1967/8 until 1973/4, the needs indicators employed were specified *a priori* and weights were assigned to them by a combined process of determining average local costs and political bargaining. From 1974/5 until 1980/1 the needs indicators were chosen as the result of a stepwise multiple regression analysis. This method attempted to 'explain' total local authority expenditure in each area in terms of 'needs' indicators as independent variables. The dependent variable in this procedure was the expenditure per head of each 'needs authority' (the county and district sums). The independent variables varied from year to year, but were consistent in including measures of population size, population characteristics, and size and density variation of local authorities. From this multiple regression the values of the regression coefficients were determined. However, the final regression formula used in grant allocation was modified in three ways in practice. First, any variable included in the original regression which yielded a negative coefficient was omitted and the regression recalculated. This represented a *positivity constraint*. This did not apply to one needs indicator, the number of low income households in an authority, which was deemed too important to omit. This variable was included in the final regression calculation, but omitted from the formula used for the distribution of the needs element. A second modification was to incorporate the effect of the total financial resources available for allocation under the needs element of RSG. This was decided exogenously to the RSG procedure by central government and represented a *global expenditure constraint.* A third modification was imposed to ensure that the needs allocation to each authority did not fluctuate too wildly from year to year. In 1975/6 this modification restricted the use of the regression formula to allocation of only 28.7% of the needs element. The remaining 71.3% was allocated solely on the basis of the needs element entitlement in the previous year (1974/5). The use of this procedure imposed a strong damping factor on needs allocation and represents a *hold-harmless constraint.*

The *resources element* accounted for about 25% of the total RSG and was designed to compensate differences between local authority rateable values (tax base). Since these control the ability to raise revenue locally by rates, a subsidy was paid to each local authority which fell below a predetermined per capita rateable value standard for England and Wales, i.e.

$$T_{t,i}^{R} = (\sigma P_{t,i} - V_{t,i})r_{t,i}$$

where $T_{t,i}^{R}$ is the finance given to local authority i under the resources element, $P_{t,i}$ is the population in authority i, $V_{t,i}$ is the total rateable value of authority i, and $r_{t,i}$ is the tax rate (or rate poundage) exacted in the authority, all at time t. The constant σ was defined by central government prior to allocation of the resources element, termed the 'standard rateable value per head'. The amount of resources element received by an authority was governed by the setting chosen for the standard rateable value per head σ. The value of σ multiplied by the population $P_{t,i}$ gives a measure of the total tax capacity of any authority based on its population, if the authority has the standard rateable value. The difference between the standard and the actual rateable value of the authority $V_{t,i}$ determined two conditions for allocation of the resources element. First, when the standard rateable value exceeded the local rate base, then a resources element was paid to that authority. The magnitude of this payment was equal to the difference in standard and actual rate bases, times the rate of local tax exacted $r_{t,i}$. The second condition applied when the standard rateable value was less than or equal to the local rate base. In this situation, no resource payment was made to the authority since it was assumed that the authority had sufficient resources of its own to cover its required expenditure. Hence a *positivity constraint* was also imposed on the resources element to ensure that, like the needs element, an allocation by central government did not result in an actual loss of finance to a local authority.

The *domestic element* represented about 15% of the total RSG and was a subsidy to domestic ratepayers by which the full poundage of their rateable values was reduced by 18½p in England and 36p in Wales; non-domestic (commercial) property owners received no subsidy. The domestic element was hypothecated, i.e. the local authorities were obliged to use this element to reduce the tax rate to domestic ratepayers. Hence, it differed from both the needs and resources elements which were block grants, and was an incongruity in the RSG. As a result, it was often suggested that it should be removed to become part of the specific grants category. As a result it is not discussed further in this chapter.

A consequence of wide-ranging criticisms directed against the RSG (see Foster *et al*. 1980, Bennett 1982a) was a severe modification of the method of grant allocation used since 1981 to a form of unitary grant combining need and resource equalisation, termed a *block grant*. This new grant takes its form from the theory of unitary grants developed by Musgrave (1961) and Boyle (1966) and is given by

$$G_i = E_i - kB_i\bar{t}(E_i - \bar{E}_i)$$

where G_i is the level of block grant, E_i is the local expenditure per head, $\bar{E}_i$ is a grant-related expenditure per head (GRE) set by central government as a standard for the area, B_i is the rateable value per head (tax base), $\bar{t}$ is the grant-related poundage (GRP) set by central government as a standard for particular classes of local authority, and k is a 'tapering' multiplier used to reduce grants to 'overspending' authorities. The GRE, or $\bar{E}_i$, is roughly equivalent to the previous needs element and is defined by applying cost weights to the size of client groups in each area. The GRP, or $\bar{t}$, when multiplied by the tax base B_i, is roughly equivalent to the previous resources element. It is defined by taking a specified national rate poundage and then dividing this between different classes of authority in terms of their differing responsibility for services. In addition there are safety nets, special arrangements for London and a number of other features. The control solution developed below adopts a form of this unitary grant framework.

7.3 *The Rate Support Grant as a control problem*

The distribution of RSG based upon the methods of optimal control has been discussed by Bennett and Tan (1979), who give a control theory solution for RSG allocation with separate grant elements (of needs, resources and domestic elements). Tan (1979a), in contrast, discusses the manner in which control-theory solutions can be applied to combined grant elements such as a unitary block grant. Empirical analyses have been developed with these methods by Tan (1979b) for the 12 Inner London Boroughs and the City of London, by Bennett and Tan (1981) for 13 Outer London Boroughs, and by Bennett (1981, 1982b) for the 115 'need' authorities in England and Wales using a hierarchical solution.

To distribute the RSG using optimal control solutions, a model is required with the structure developed in earlier chapters:

$$Y_t = A_t Y_{t-1} + B_t X_t + b_t \tag{7.1}$$

where Y_t is a $n^* \times 1$ state vector, X_t is a $m^* \times 1$ control vector, b_t is a $n^* \times 1$ vector of known exogenous variables not subject to control. The matrices A_t and B_t are coefficient matrices of dimension $n^* \times n^*$ and $n^* \times m^*$ respectively. The above model may be either assumed known *a priori*, or estimated from available data. The precise structure of the dynamic model (7.1) depends upon the form of grant allocation procedure adopted and will differ between countries. For a block grant structure, similar to that of RSG which existed after the 1981/2 financial year, Equation (7.1) is specified as follows:

$$\begin{bmatrix} E_{ti} \\ R_{ti} \end{bmatrix} = \begin{bmatrix} \alpha_{1i} & 0 \\ 0 & 1 \end{bmatrix} \begin{bmatrix} E_{t-1,i} \\ R_{t-1,i} \end{bmatrix} + \begin{bmatrix} \alpha_{3i} & \alpha_{2i} V_{ti} \\ 0 & V_{ti} \end{bmatrix} \begin{bmatrix} T_{ti} \\ \Delta r_{ti} \end{bmatrix} + \begin{bmatrix} \alpha_{0i} \\ 0 \end{bmatrix} \tag{7.2}$$

where E_{ti} is the total expenditure, R_{ti} is the total rate yield (local tax income), V_{ti} is the rateable value (tax base), T_{ti} is the grant allocation received, Δr_{ti} is the change in rate poundage from $t-1$ to t (change in tax rate), and α_{ji} are the parameters to be determined.

Each term is defined for local authority i at time t. This equation is made up of two functional expressions:

(a) $E_{ti} = \alpha_{1i} E_{t-1,i} + \alpha_{2i} V_{ti} \Delta r_{ti} + \alpha_{3i} T_{ti} + \alpha_{0i}$

This expresses the dynamic dependence of expenditures on past expenditures, rate yields and grants and is a supply equation.

(b) $R_{ti} = R_{t-1,i} + V_{ti} \Delta r_{ti}$

This is definitional, stating that rate yield at each time period is equal to previous rate yield plus change in rate poundage multiplied by rateable value.

Equation (7.2) is a special form of (7.1) in which we let

$$A_{ti} = \begin{bmatrix} \alpha_{1i} & 0 \\ 0 & 1 \end{bmatrix} \quad \text{and} \quad B_{ti} = \begin{bmatrix} \alpha_{3i} & \alpha_{2i}V_{ti} \\ 0 & V_{ti} \end{bmatrix}$$

As there are N authorities, there will be one equation for each authority. These can be reduced to a single equation of the form (7.1) by making the following definitions:

$$A_t = \begin{bmatrix} A_{t,1} & 0 & \dots & 0 \\ 0 & A_{t,2} & \dots & 0 \\ \vdots & \vdots & & \vdots \\ 0 & 0 & \dots & A_{t,N} \end{bmatrix},$$

$$B_t = \begin{bmatrix} B_{t,1} & 0 & \dots & 0 \\ 0 & B_{t,2} & \dots & 0 \\ \vdots & \vdots & & \vdots \\ 0 & 0 & \dots & B_{t,N} \end{bmatrix} \tag{7.3}$$

$$Y_t = \begin{bmatrix} Y_{t1} \\ Y_{t2} \\ \vdots \\ Y_{tN} \end{bmatrix}, \quad X_t = \begin{bmatrix} X_{t1} \\ X_{t2} \\ \vdots \\ X_{tN} \end{bmatrix}, \quad b_t = \begin{bmatrix} b_{t1} \\ b_{t2} \\ \vdots \\ b_{tN} \end{bmatrix}$$

where

$$Y_{ti} = \begin{bmatrix} E_{ti} \\ R_{ti} \end{bmatrix}, \quad X_{ti} = \begin{bmatrix} T_{ti} \\ \Delta r_{ti} \end{bmatrix} \quad \text{and} \quad b_{ti} = \begin{bmatrix} \alpha_{0i} \\ 0 \end{bmatrix}$$

The block diagonal structure of the matrices A_t and B_t in (7.3) assumes, as a first approximation, that each of the N authorities is uncoupled from the rest. Other forms of expenditure model can also be defined (see, for example, Eastwood 1978, Bennett 1982a).

As discussed earlier, under the RSG, a predetermined limit is set by the central government for the total amount of grant available for distribution to the local authorities. This imposes a linear equality constraint on the control variable T_t. As each X_{ti} is made up of T_{ti} and Δr_{ti}, it is clear that if this cash limit for time t is denoted by f_t then the constraint may be written as

$$[1, 0, 1, 0, \ldots, 1, 0] \begin{bmatrix} T_{t1} \\ \Delta r_{t1} \\ T_{t2} \\ \Delta r_{t2} \\ \vdots \\ T_{tN} \\ \Delta r_{tN} \end{bmatrix} = f_t$$

which is a specific case of the more general form of linear equality constraint

$$F_t X_t = f_t \tag{7.4a}$$

A second restriction on grant allocation is the unidirectional flow of finance – only from higher to lower levels of government. This means that T_{ti} must be positive for each authority and for each time period. Such a constraint may be specified as

$$R_t X_t \geqslant 0 \qquad \text{for each } t \tag{7.4b}$$

and where the matrix R_t is defined as

$$R_t = \begin{bmatrix} 1 & & & & & & \\ & 0 & & & & \mathbf{0} & \\ & & 1 & & & & \\ & & & 0 & & & \\ & & & & \ddots & & \\ & \mathbf{0} & & & & 1 & \\ & & & & & & 0 \end{bmatrix} \tag{7.5}$$

These two constraints are in the form used in previous chapters. However, note that with R_t given according to (7.5), the constraint (7.4b) is 'weaker' than the constraint $X_t \geqslant 0$ in the sense that the latter also places a restriction on the rate poundage.

The control theory approach to the problem of allocating the RSG proceeds by using the state variables in Equation (7.1) to define equalisation criteria which are used as target levels which it is sought, within the RSG allocation procedure, to match. These targets will be levels of both expenditure matching and tax yield (rate poundage multiplied by tax base) matching. Having defined target levels, as a_t, the control approach seeks to achieve that level of grant allocation which renders a minimum the difference between targets and the actual level of expenditure requirements; i.e. achieves the fullest possible equalisation by making the term $(Y_t - a_t)$ as small as possible. The control approach allows these differences to be summed over the planning time horizon adopted. The time horizon adopted in the present example is a five-year one which, with annual data, gives a five-stage optimisation problem. The multistage nature of the optimisation problem is important since it allows trade-offs in the future for dynamic changes in state in the present; vice versa, it allows trade-offs in the present for dynamic changes anticipated in the future. An important example where such trade-offs become important is in city growth or decline where expenditure on public goods, such as schools or hospitals, can take account of changing expenditure needs resulting from changed population levels. This is important in improving grant certainty and stability.

The solution developed below assumes the parameter matrices A and B to be unknown but constant. The control problem is then to find that set of control variables which minimise the quadratic function

$$V = E_0(W) = E_0\left[\sum_{t=1}^{T} (Y_t - a_t)'M_t(Y_t - a_t)\right] \tag{7.6}$$

subject to the linear equality and inequality constraints (7.4a) and (7.4b).

The strategy for solving this inequality-constrained problem is exactly that developed in previous chapters and is composed of two parts. First, the control problem is solved with only the equality constraints imposed. Second, if the resulting solution is not feasible in the sense that there exists at least one local authority for which the inequality constraint is not satisfied, then the grant allocations need to be adjusted for feasibility using the dual simplex method discussed in Section 3.3.

7.4 Control theory solutions

In the following discussion two examples are developed which demonstrate the use of this control method for grant allocation in small scale and large scale systems. The first case utilises the control theory results for a 13-region problem of grant allocation within Inner London. The second case seeks optimal control results for allocation to 115 local authorities in England and Wales which, because of the high dimension of the resulting control matrices, necessitate a hierarchical solution strategy.

Example 1: Grant allocation in Inner London Empirical results of applying the optimal control algorithm discussed above are developed in this section for 13 of the Inner London boroughs for the five years from 1974/5 to 1978/9. This is a relatively small-scale investigation, but evidences the general properties of the control results.

Three assumptions are required prior to the start of the analysis. First, the target values a_t have to be specified. These are not available from the RSG allocation methods actually employed in Britain, and in the present analysis the target values of the state variables are set equal to their estimated values given in published statistics. A second assumption relates to the choice of weights M_t to enter the objective function. Again these are not available from the actual RSG allocation procedure. Those used below are derived by taking into account the varying magnitude of each state variable, i.e.

$$m_{iit} = \left(\frac{\text{value of largest target variable}}{\text{target value of variable of } i\text{th authority}} \right)^2 \qquad (7.7)$$

Using this method, the weights will vary as a function of both time and location and deviations from targets will each be adjusted to the same scale at any location. A third assumption relates to the initial conditions. These are assumed to derive from the historic past for the fiscal year 1973/4.

Using these assumptions the resulting control theory solution follows three stages: estimation of the state equation; derivation of the equality-constrained control solution; and derivation of the inequality-constrained solution.

Table 7.1 Estimated coefficients for state equation (7.2).

Borough	α_0	α_1	α_2	α_3	R^2
City of London	13.3500	0.9224	0.8430	−0.8202	0.9840
Camden	11.5920	0.6698	0.8288	1.2369	0.9984
Greenwich	8.8533	0.4913	−0.2349	0.7487	0.9677
Hackney	8.6080	0.2027	0.9889	1.2905	0.9937
Hammersmith	7.4814	0.2196	0.9154	1.3097	0.9864
Islington	12.2510	0.3035	0.9242	1.3205	0.9921
Kensington and Chelsea	6.0194	0.4293	0.9967	1.5725	0.9999
Lambeth	1.5612	0.2255	1.5644	1.6998	0.9999
Lewisham	1.8169	0.6319	−0.3774	0.7010	0.9248
Southwark	15.4491	0.3118	0.6545	1.1347	0.9809
Tower Hamlets	9.3327	0.1272	0.8609	1.7748	1.0000
Wandsworth	1.3743	0.6878	−0.0290	0.7754	0.9226
Westminster	32.2767	0.7638	0.7355	1.2874	0.9887

ESTIMATION OF THE STATE EQUATION Ideally this problem should be tackled with relatively sophisticated techniques such as recursive filters. However, with the present five-stage control problem, the shortness of the available data preclude such an approach. Instead, the estimates of the regression parameters are obtained below using OLS. In this context OLS can be interpreted only as an averaging tool, and certainly statistical significance tests cannot be employed. The resulting estimates of the parameter values are given in Table 7.1. These show, in general, a predictable pattern. The response of local expenditure to grants (α_{3i}) gives a generally high and positive elasticity. Similarly the response to past expenditures (α_{1i}) is also positive, but with a lower value.

EQUALITY-CONSTRAINED SOLUTION A second stage of analysis is the solution to the control problem with the constraint (7.4a) imposed, but ignoring (7.4b). The results of this calculation are given in Table 7.2. There are two control variables: the central grant, and the local rate poundage. Table 7.2 demonstrates that, without inequality constraints imposed to prevent removal of grants from local areas, one area, the City of London, pays a considerable contribution to the grants of other areas. This results both from its high tax base and its negative expenditure elasticity estimate, as shown in Table 7.1. The lack of inequality constraint also means that the City pays the bulk of the historic rise in local expenditures.

INEQUALITY-CONSTRAINED SOLUTION A final stage of analysis is the solution to the control problem with both constraints (7.4a) and (7.4b) imposed. The results of this calculation are given in Table 7.3. In comparison with the equality-constrained case of Table 7.2 this shows a much more even level of grant distribution between the local boroughs, and a more general rise in rate poundage overall. The City changes from being a large net contributor to a position in which it receives no grants and has to raise its rate poundage fairly rapidly.

A comparison of the control solutions with the actual or desired allocations is shown in Figures 7.1–7.3. These figures display the four variables of interest – grant allocated T_t, rate poundage r_t, expenditure E_t and rate yield R_t. Each figure displays thirteen graphs, one for each Inner London borough. For the two control variables Figures 7.1 and 7.2 show that there is generally more variation for the grant control variable than for the rate poundage variable. However, in both cases the results for the first period show very small differences between the calculated and actual values, although the variation increases with

Table 7.2 Optimal grants and rate poundages in equality constrained case (7.4a).

Borough	*Grants (£m)*					*Rate poundage (p in £)*				
	1974/5	*1975/6*	*1976/7*	*1977/8*	*1978/9*	*1974/5*	*1975/6*	*1976/7*	*1977/8*	*1978/9*
City of London	0.6	−35.1	−22.9	−17.7	−0.8	41.2	61.9	66.8	68.9	70.0
Camden	10.1	21.5	28.9	24.9	33.9	53.5	78.5	88.0	99.6	95.6
Greenwich	11.9	23.8	27.8	26.0	26.0	51.4	71.1	74.1	81.7	94.4
Hackney	13.0	18.2	28.4	28.6	31.2	46.9	64.6	68.7	71.6	65.5
Hammersmith	9.3	16.1	24.0	22.3	24.0	48.6	69.3	69.1	80.9	84.4
Islington	12.0	20.2	29.7	27.2	29.6	51.2	74.6	74.8	82.7	83.4
Kensington and Chelsea	9.4	15.4	22.0	19.3	22.7	41.1	59.5	61.9	70.0	69.3
Lambeth	15.6	22.4	34.4	33.5	38.9	52.7	71.6	76.0	84.1	81.5
Lewisham	18.8	28.9	29.4	29.6	29.8	51.4	69.1	71.0	79.0	97.8
Southwark	14.7	24.0	38.5	35.6	38.1	54.9	75.9	83.3	88.3	86.7
Tower Hamlets	11.2	15.3	22.8	22.6	24.1	59.2	79.8	81.2	88.6	88.1
Wandsworth	20.6	26.9	32.4	28.6	25.7	47.1	64.8	69.0	80.3	100.6
Westminster	5.8	30.3	41.6	26.3	56.2	45.0	67.5	75.1	81.0	75.7

Table 7.3 Solution to control problem with equality and inequality constraints (7.4a) and (7.4b).

	Grants (£m)					Rate poundage (p in £)				
Borough	*1974/5*	*1975/6*	*1976/7*	*1977/8*	*1978/9*	*1974/5*	*1975/6*	*1976/7*	*1977/8*	*1978/9*
City of London	0.0	0.0	0.0	0.0	0.0	39.4	66.7	76.4	81.3	82.2
Camden	10.7	17.5	26.6	22.6	34.0	52.0	79.0	89.5	101.9	97.7
Greenwich	11.2	22.8	26.8	25.1	25.8	51.6	71.0	74.0	81.4	94.1
Hackney	12.3	17.1	27.3	27.8	30.9	46.4	64.7	69.1	72.1	65.9
Hammersmith	8.8	15.2	23.2	21.6	23.7	48.2	69.4	69.5	81.4	84.9
Islington	11.3	18.4	28.5	26.2	29.3	50.4	74.9	75.4	83.6	84.2
Kensington and Chelsea	9.2	14.3	21.3	18.7	22.6	40.5	59.7	62.3	70.7	70.0
Lambeth	15.2	20.4	32.9	32.3	38.6	51.6	71.9	76.8	85.4	87.6
Lewisham	18.5	27.7	28.5	28.6	29.7	51.8	69.0	70.7	78.6	97.2
Southwark	13.2	22.0	36.2	34.0	37.6	54.3	76.1	83.8	89.1	87.4
Tower Hamlets	10.6	14.5	22.1	22.0	23.9	58.6	79.9	81.6	89.2	88.7
Wandsworth	20.3	25.6	31.2	27.4	25.8	47.1	64.8	69.0	80.3	100.5
Westminster	11.8	12.4	32.6	20.2	57.6	42.2	68.5	77.8	84.8	79.0

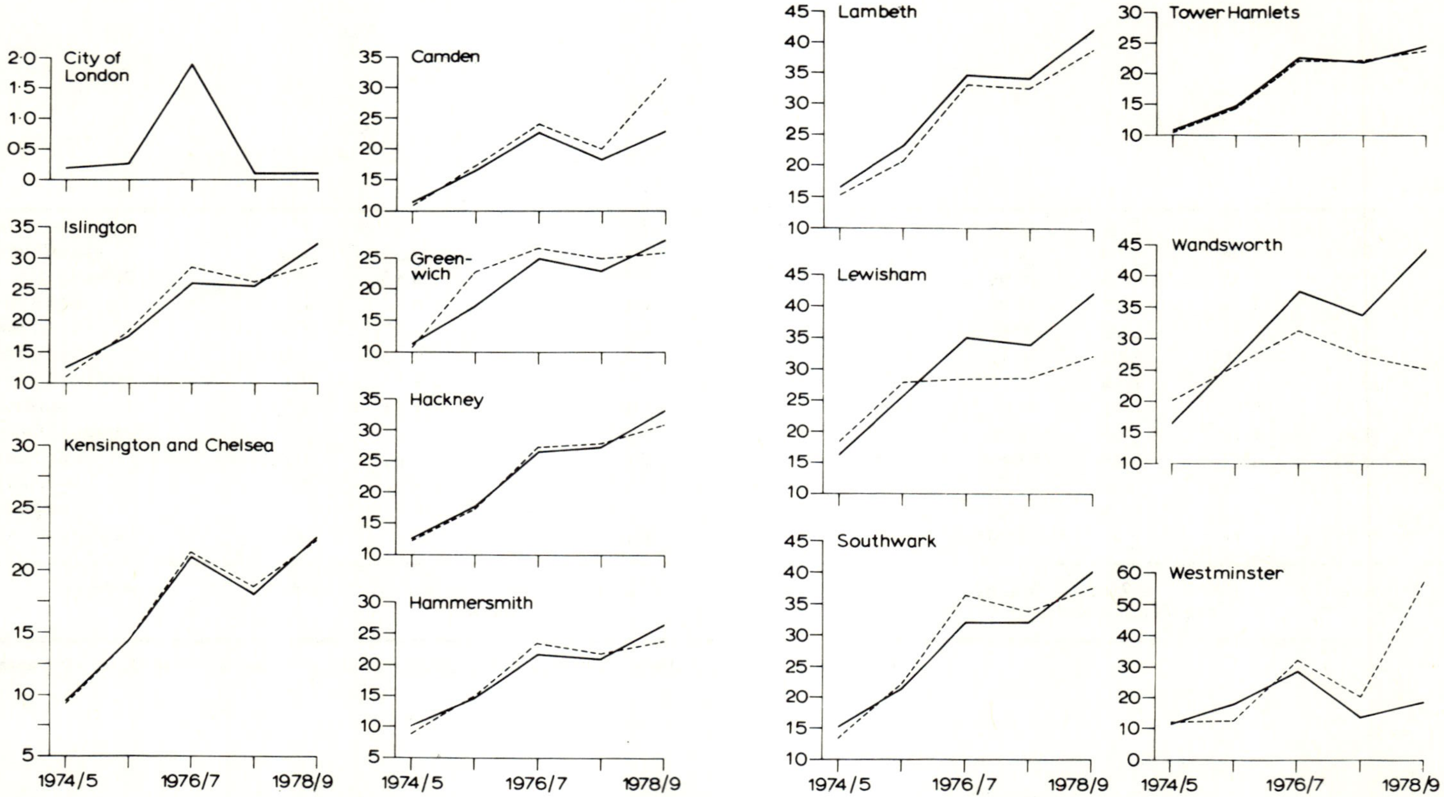

Figure 7.1 Comparison of control solutions (dashed lines) and actual allocations (solid lines) of Rate Support Grant (£m).

time. Also, observe that the calculated optimal local tax rates are consistently higher than the actual tax rates levied.

For the state variables shown for expenditure in Figure 7.3 the calculated values for R_t track the desired target trajectory of R_t better than for the E_t variable. As with the control variables, the variation from target values increases with time for both variables. Also, except for Lewisham and to some extent Wandsworth, the computed expenditure variable E_t does show a consistent underestimate when compared with the target value of E_t.

Example 2: Grant allocation in England and Wales – An hierarchical solution The second example developed here is the same as the first example: that of determining an optimal setting of intergovernmental grants and local tax rates. However, it requires a modified control approach because the dimension of the resulting control equations precludes direct solution. This example uses the 115 local authorities in England and Wales at the level of the so-called 'needs authorities' and a five-stage solution is employed for the period from 1975/6 to 1979/80 with 1974/5 as the initial condition. The weights for M_t in the objective function are the same as in the previous example and OLS is again used to estimate the state equation. However, the targets are now defined exogenously using a representative index of expenditure need tabulated by Bennett (1982a, Ch. 6) which disaggregates local spending requirements into 30 subcategories, applies client group measures to average costs in each subcategory in different types of administrative area, and then re-aggregates to give a measure of overall spending requirement for each local authority. Using this method, the targets will vary as a function of both time and location and are each adjusted to the same scale at any location.

HIERARCHICAL CONTROL SOLUTION It is frequently impossible to solve Equations (3.20)–(3.26) for an entire system of local areas simultaneously, given the constraints of computer time or central processor capacity. Various methods of overcoming this difficulty are possible. For example, the T-stage problem can be restated as a set of T separate one-stage problems. Alternatively, the N spatial zones can be disaggregated into a set of relatively independent regions and allocations solved separately for each. Another approach is to find some more approximate solution which does not require full integration of the Ricatti equations (3.26). A possible approach to this is to limit the generality of the matrices A, B or M_t. However, each of these approaches will give a poor approximation to the optimal solution or will give a considerable loss of generality.

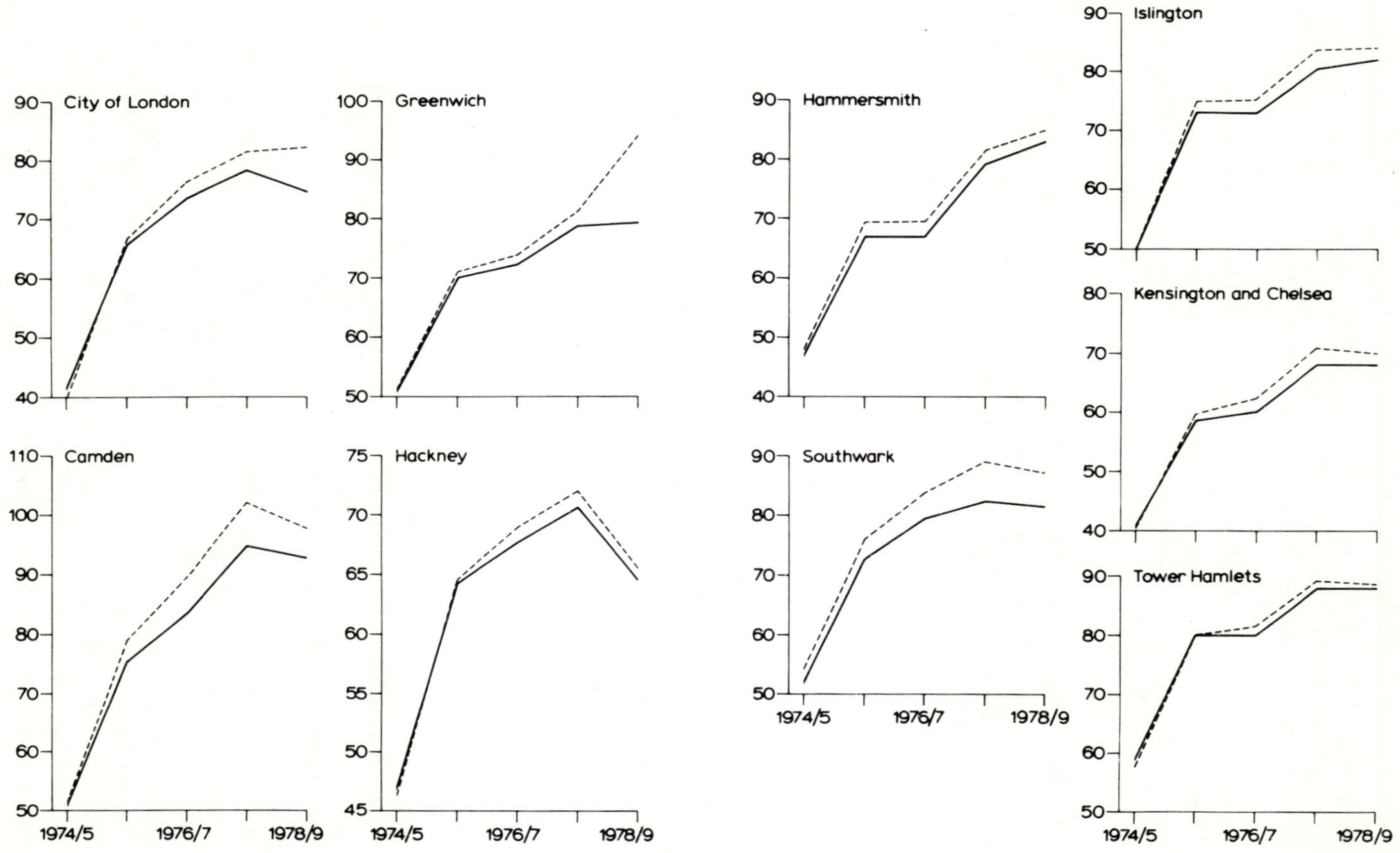
City of London
Greenwich
Hammersmith
Islington
Camden
Hackney
Southwark
Kensington and Chelsea
Tower Hamlets
1974/5
1976/7
1978/9

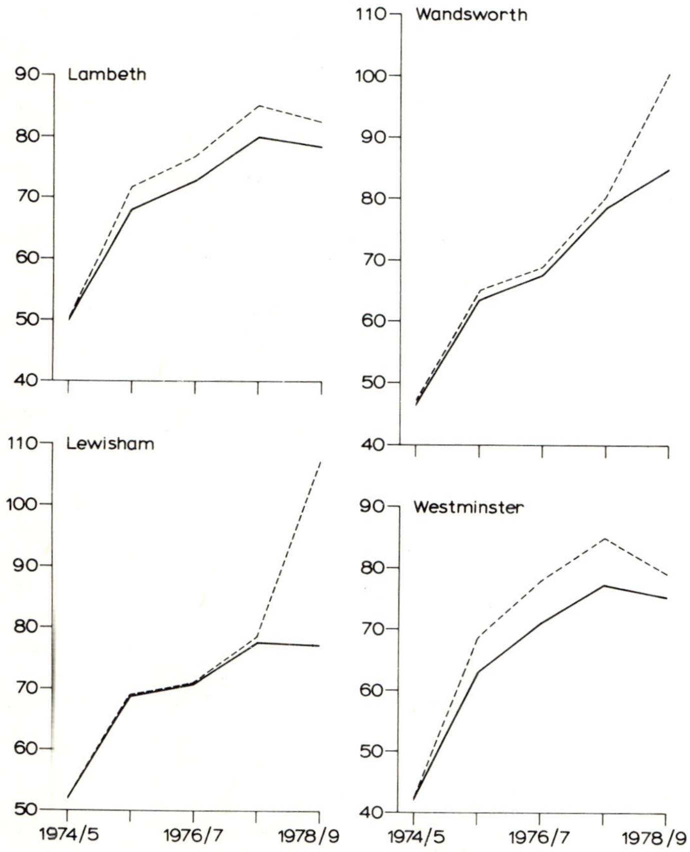

Figure 7.2 Comparison of actual local tax rates (solid lines) – rate poundage – with control solutions (dashed lines) given in pence in the pound sterling.

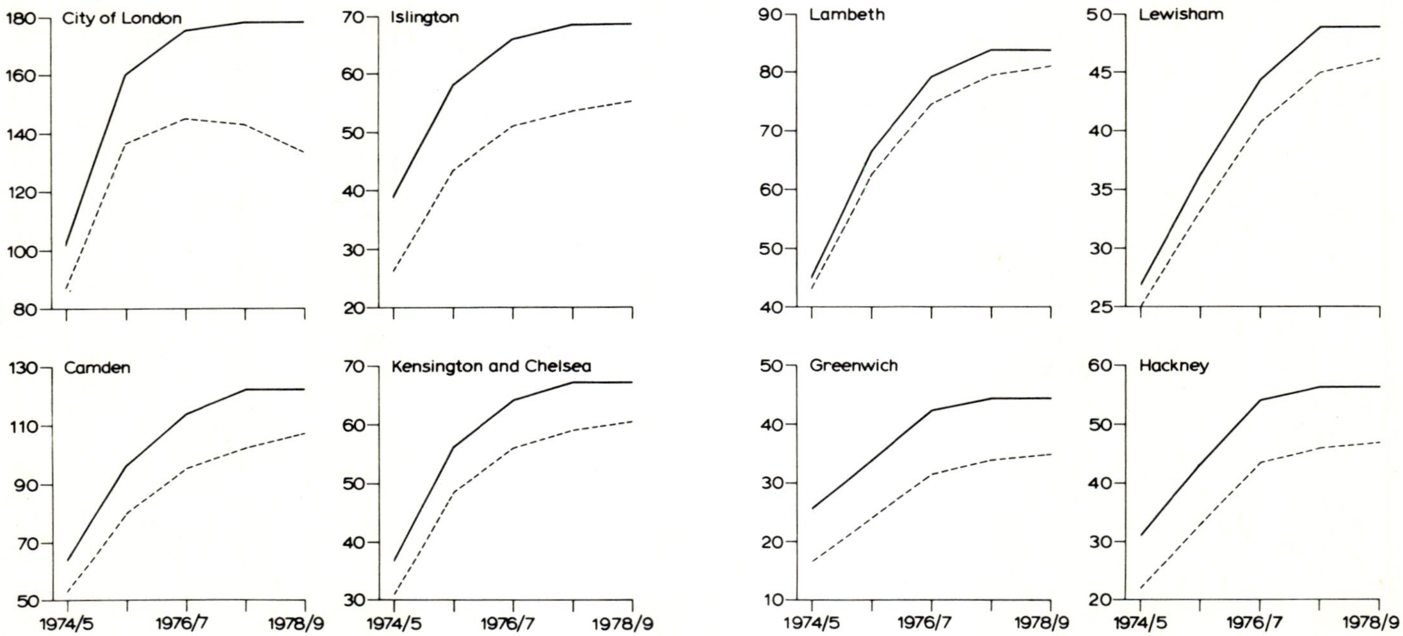
City of London
Islington
Lambeth
Lewisham
Camden
Kensington and Chelsea
Greenwich
Hackney
1974/5
1976/7
1978/9

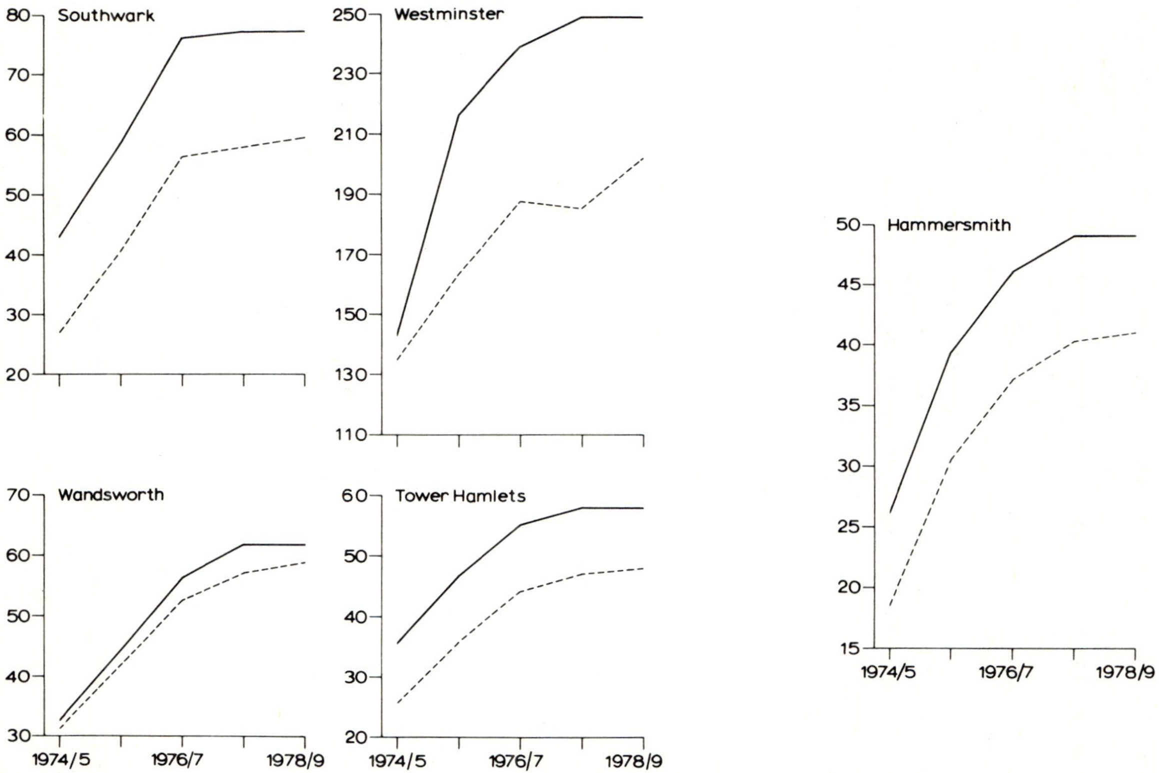

Figure 7.3 Comparison of expenditure targets (solid lines) and control theory solutions (dashed lines) (£m).

The present example uses an hierarchical solution to the dimensionality problem. This has been previously used by Bennett (1981, 1982b). This breaks up the set of N spatial regions into p smaller sets (preferably internally related), solves the resulting p separate control problems, then takes the results of these p control problems as the input into a final control solution for each of the p subsets. There are a number of strategies by which one can approach this hierarchical optimisation problem (see, for example, Schoeffler 1971, Singh 1981, Bernussou & Titli 1982). The present example proceeds with a form of what Schoeffler terms 'influence decomposition'. This involves partitioning the control variables into spatial subsets. This approach is particularly attractive in the present example, since the inequality constraints can be used, to a first approximation, to introduce the hierarchical solution. This strategy for solving the control problem in hierarchical form has two main attractive features: first, it reduces dimension and hence economises on computer resources; secondly, it is a natural solution when a multilevel intergovernmental allocation problem is involved (e.g. from federal to state and then to local level).

Using this hierarchical approach the resulting control theory solution follows two stages: (1) partitioning of the 115 local authorities into p subsets and then for each subset (a) estimate the state equation and (b) derive the inequality constrained solution; (2) enter these solutions as constraints into the second stage optimisation problem, repeating steps (1) for each subset:

Stage 1 *Partition the space and solve for a reduced spatial set*

Step (1) Split the n spatial areas of the study region into p subsets.

The 115 local authorities in England and Wales used are those referred to as the 'needs authorities', with the Isles of Scilly and the City of London omitted. These authorities fall into natural sets of the non-metropolitan counties, metropolitan districts, Inner London boroughs and Outer London boroughs. However, it was necessary to further partition these subsets in order to make the solution computationally feasible.

Step (2) Define aggregate targets and equations of state (Eq. 7.2) for each of the p subsets. The targets in this approach are given from tabulations in Bennett (1982a, Appendix 2). The state equation is estimated, as in the previous example, by OLS.

Step (3) Determine the optimal control allocation between the p subsets in the equality-constrained case using Equations (3.20)–(3.26).

Step (4) Check that the non-negativity constraint is satisfied; if it is not then follow the three-step procedure outlined above (Section 3.3).

Stage 2

Step (1) Take the output of Stage 1 to define the global resources constraints and then solve the equality-constrained control allocation using Equations (3.20)–(3.26) for first subset of areas.

Step (2) Check that the non-negativity constraint is satisfied; if it is not, then follow the three-step procedure discussed above.

Step (3) Repeat steps (1)–(2) for each of the p subsets of spatial areas.

The scale of the results of this analysis preclude complete reporting here; however, they are tabulated in full in Bennett (1981, 1982b). What is possible here is a comparison of the allocation produced from the optimal control theory solution with those which were actually made in practice. This comparison is indicated in Figure 7.4. For every area, the control theory solution gives a smaller grant allocation than the actual allocation. This is a result as much of the unitary grant feature as the control solution: the overall grant level can be reduced by taking advantage of coincidences of low resources and low expenditure need, and of high resources and high expenditure need. However, the pattern shown in Figure 7.4 also displays the marked effect of the optimal control results. In general the peripheral areas, especially the metropolitan districts and central cities, but also parts of Inner London, lose the largest amount of grant. The optimal allocation solution is directing grant away from city areas which, although they may have high spending need, nevertheless have high local tax resources. Similarly, the optimal control solution directs grants away from areas such as Wales which, although it has very poor local tax base, has also very small local need to spend. The aggregate pattern is one which differs greatly from that which actually occurred in 1978/9 and, had it been implemented, would have caused considerable controversy.

7.5 Conclusion

This chapter has presented the results of applying a hierarchical formulation of optimal control theory to a rolling five-year programme for allocation of the Rate Support Grant in Britain. These are major full scale practical applications of the constrained control algorithm developed in earlier chapters, and exhibit the potentiality of the method. Other similar problems which could be tackled by this

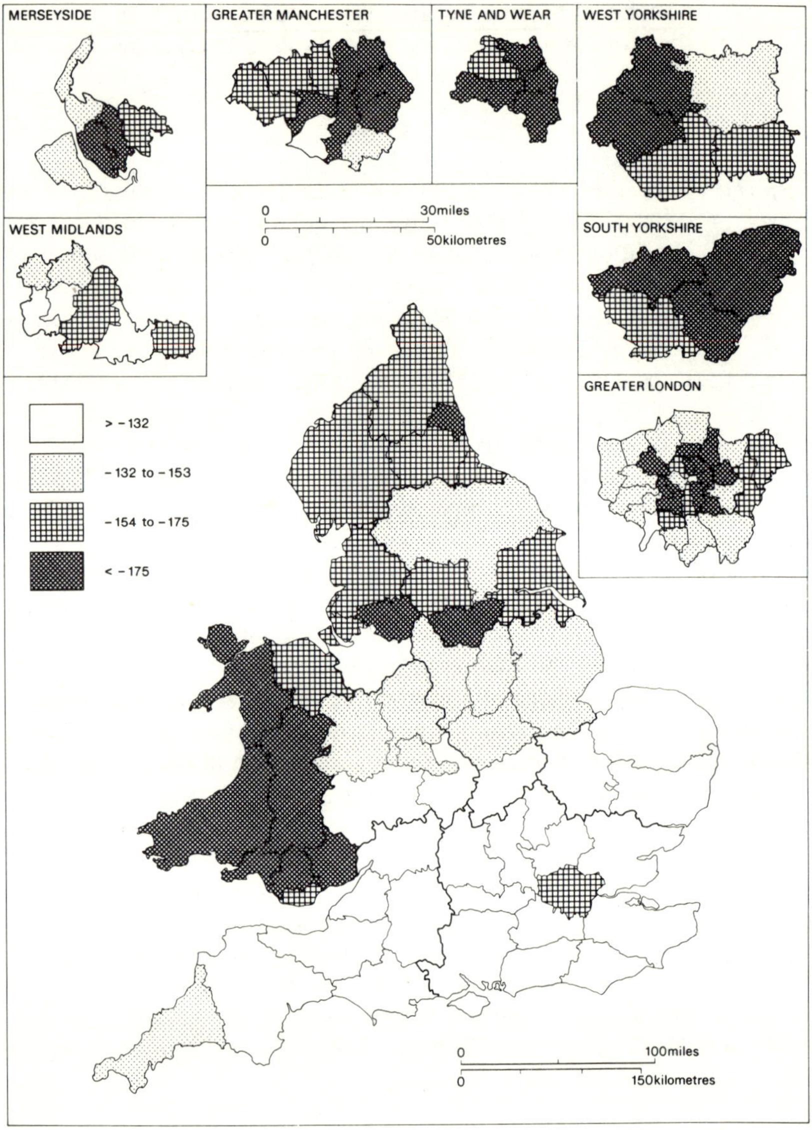

Figure 7.4 Deviation of control theory solutions for allocation of Rate Support Grant from actual allocations in 1978/9 (in pounds sterling per head). (*Source*: Bennett 1982b.)

approach is the allocation of revenue-sharing grants as a three-level problem from federal to state and local governments in the USA; the Länder and Gemeinden equalisation grants used in Germany; and the interprovince, state and local government grants used in most Western countries.

8 *Planning applications of control theory: national settlement systems and manpower*

8.1 Introduction

The planning process has been one of the richest areas for generating control problems with spatial systems. Here the decision process is to seek control of land use, industrial location, population movement, the location of retail facilities, and employment change. Early stimulus to this work was given by Chadwick (1971) and McLoughlin (1973) and more recently this has been given an important further stimulus by a movement towards a more continuous pattern of planning. In such 'continuous planning' the evolution of events is kept under continuous review and indicators of change in spatial outcome are compared against various indications of plan performance. This necessitates the construction of an information and monitoring system which can track changes and deviations from targets. Such a planning-monitoring configuration is shown in Figure 8.1. Here the range of deviations from planning targets is used to trigger a series of possible planning responses. These span a set of decisions from 'no action' (i.e. accommodation to change) to complete plan review where deviations have been so great that the targets must be redrawn. The most common actions to be found in most planning-control systems, however, are found between the two extremes: as implementation and impact decisions.

Most of the developments in planning derive either from the work of economists on macro-economic control, or from transport engineering on optimal network layout and vehicle scheduling algorithms. More recently this corpus of work has been expanded by the tackling of planning problems using control frameworks which use explicit spatial information and an explicit spatial model structure. In the context of *environmental planning*, tackling such problems as pollution control, flood routing, etc., the literature is summarised by Bennett and Chorley (1978) and Wilson (1981). Major examples are the studies by Nijkamp (1975, 1978, 1979a,b), Sakawa and Ueda (1978), Whitehead *et al.* (1978) and Whitehead and Young (1979). In *transport planning* Ralston (1979) and Barber and Ralston (1980) develop a set of

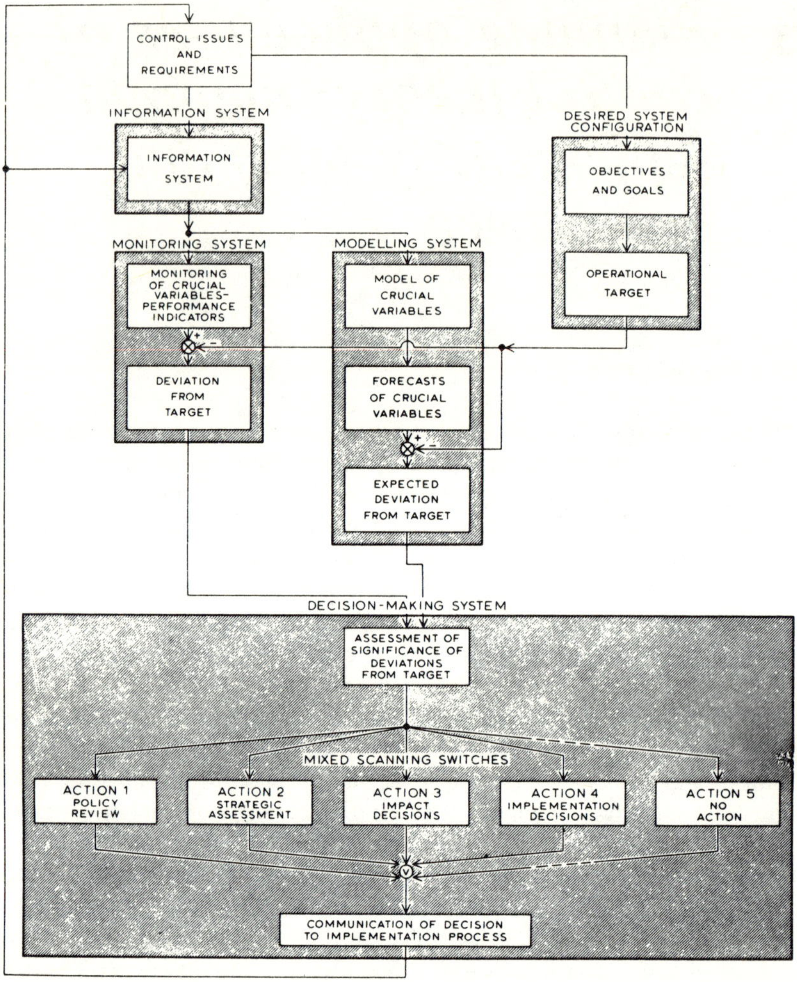

Figure 8.1 The interaction of monitoring, advisory systems, information systems and control solutions as a mixed scanning process of possible control responses (after Bennett & Chorley 1978).

dynamic route-planning algorithms. In planning the location of *retail facilities*, Phiri (1979, 1980) provides a good review of the literature. *Land-use planning* control of the land market has been tackled as a spatial control problem in a number of papers which are reviewed in Shoup (1970), Ripper and Varaiya (1974), Anas (1978), Foldes (1978a,b) and Fujita (1980, 1981). For *health-care planning* a seminal analysis for Britain was undertaken by West (1973, 1979) and McDonald *et al.* (1974) which has been developed by Le Grand (1976, 1978). Other work on this problem is that by Leonardi (1978), and

Mayhew and Leonardi (1982) (see also Wilson 1981, Ch. 9). For *regional economic planning* the work of Raman (1963), Sakashita (1967), Domanski (1973), Tabata (1976), Snickars (1978), Sakashita and Kamoike (1973) and Lee (1982) are some of the main sources. Similar results have been developed for *education planning* by Felderer (1975).

The application of control approaches to planning has not been without its problems, as is the case with the rest of planning practice. Some of the more general of these are outlined by Hall (1974, 1980), while Batty (1971, 1974, 1976) has approached the problem of how control models in planning can be meshed with the participation process and optimal estimates of targets obtained by forms of collective action. In a wider methodological context, Gregory (1980) has castigated control frameworks as largely 'instrumentalist', leading to mechanisms not only of planning regulation, but also opening up opportunities for social control and social repression. Such dangers are inherent in any governmental activity and must be strongly guarded against in the implementations of the techniques discussed here. Below we develop two examples of planning control theory; first, for planning population levels in a national system of settlements, and secondly, for planning of sectoral manpower requirements in a national economy.

8.2 *Two national settlement system planning models*

There is now a large body of research concerned with using urban and regional planning models to aid decision making in the planning process. Most commonly such models have been used for decision on the design of transport networks, the location of retailing outlets or the location of public facilities (see, for example, Lowry 1964, Wilson 1974). Recently, however, considerable attention has also been given to the population growth and migration component of such models and this has given rise to the issue of national settlement system planning. It is this example which is developed at length below, but the discussion is applicable to the whole range of urban and regional planning models which have been developed in a dynamic framework.

There are two main forms of national settlement system planning model which have been developed: the forward-linkage and backward-linkage models (Tan 1980). The *forward-linkage model* is a direct extension of the population cohort-survival model developed by Keyfitz (1968), Rogers (1968) and Rees and Wilson (1977). In its simplest form the forward-linkage model can be stated as follows:

$$\tilde{Y}_t = (N + M)\tilde{Y}_{t-1} + x_t \tag{8.1}$$

where $\tilde{Y}_t$ is the population distribution vector in age cohorts at time t, x_t represents the number of people migrating to each region from outside the system in each age cohort; N is a diagonal matrix of natural rates of increase and M is a matrix of transition probabilities (the (i, j)th element is the probability that a person currently residing in region j will migrate to region i during one time period), where

$$N = \begin{bmatrix} N_{11} & 0 & \dots & 0 \\ 0 & N_{22} & \dots & 0 \\ \vdots & \vdots & & \vdots \\ 0 & 0 & \dots & N_{nn} \end{bmatrix}, \; M = \begin{bmatrix} M_{11} & M_{12} & \dots & M_{1n} \\ M_{21} & M_{22} & \dots & M_{2n} \\ \vdots & \vdots & & \vdots \\ M_{n1} & M_{n2} & \dots & M_{nn} \end{bmatrix}$$

and

$$N_{ij} = \begin{bmatrix} 0 & \dots & 0 & \alpha_{1i} & \alpha_{2i} & \dots & \alpha_{qi} & 0 & \dots & 0 \\ 0 & \dots & 0 & 0 & 0 & \dots & 0 & 0 & \dots & 0 \\ \vdots & & \vdots & \vdots & \vdots & & \vdots & \vdots & & \vdots \\ 0 & \dots & 0 & 0 & 0 & \dots & 0 & 0 & \dots & 0 \end{bmatrix}$$

where α_{ji} are age-specific birth rates over the q cohorts containing the child-bearing years.

The *backward-linkage model* assumes that the introduction of specific temporal and spatial stimuli will enable the system to be steered as closely as possible towards the desired population distribution level. For example, assuming that job vacancies are the stimulus to which migrants respond (Evtushenko & MacKinnon 1976), the backward-linkage model may be represented as

$$\tilde{Y}_t = (I + N)\tilde{Y}_{t-1} + (I - \bar{M})v_{t-1} \tag{8.2a}$$

$$v_t = \bar{M}v_{t-1} + x_t + \delta_t \tag{8.2b}$$

where v_t is the distribution of job vacancies in period t, x_t is the distribution of government-stimulated job vacancies, δ_t is the distribution of spontaneously occurring vacancies and $\bar{M}$ is the migration matrix (the (i, j)th element is the probability that a person living in region i will fill a job vacancy in region j). In both models the control vector is denoted by x_t.

8.3 *National settlement system planning as a control problem*

The backward-linkage model represented by (8.2) is inconvenient for optimal control calculations. By defining a new augmented state vector y_t, where

$$y_t = \begin{bmatrix} \tilde{Y}_t \\ v_t \end{bmatrix} \tag{8.3}$$

the model (8.2) may be rewritten as

$$y_t = \begin{bmatrix} I + N & I - \bar{M} \\ 0 & \bar{M} \end{bmatrix} y_{t-1} + \begin{bmatrix} 0 \\ I \end{bmatrix} x_t + \begin{bmatrix} 0 \\ I \end{bmatrix} \delta_t \tag{8.4}$$

which is the familiar state-space form given by (2.13).

The constraints which are imposed on the controls x_t can be represented in the general form

$$F_t x_t = f_t \tag{8.5a}$$

$$\tilde{R}_t x_t \leqslant \tilde{r}_t \tag{8.5b}$$

$$x_t \geqslant 0 \tag{8.5c}$$

for all $t = 1, 2, \ldots, T$. The global constraint for the entire planning period is given by

$$\tilde{S}_1 x_1 + \tilde{S}_2 x_2 + \ldots + \tilde{S}_T x_t \leqslant \tilde{s} \tag{8.6}$$

If x_t is a $m \times 1$ control vector, then in (8.5) F_t and $\tilde{R}_t$ are known matrices of dimension $q_1 \times m$ and $q_2 \times m$ respectively, and f_t and $\tilde{r}_t$ are vectors of known quantities of dimension $q_1 \times 1$ and $q_2 \times 1$ respectively. It is assumed that the rank of F_t is equal to q_1 and is less than m. For (8.6), each $\tilde{s}_i$ is a $q \times m$ known matrix and $\tilde{s}$ is a $q \times 1$ known vector.

In the forward-linkage model x_t denotes migration to each region from outside the system. Thus in (8.5a) F_t is a two-row vector with all elements equal to unity and f_t represents the total immigration pool available in time t. The constraint of (8.5b) can be interpreted as a budget constraint for time t with $\tilde{R}_t$ a row vector the ith component of which represents the cost of encouraging migration of a person to region i, and $\tilde{r}_t$ is the budget available at time t. The constraint (8.6) may be interpreted in the same manner to represent a budget constraint for the entire planning period.

Let the vector $\tilde{a}_t$ denote the desired population distribution at time t. The trajectory of the desired population distribution level is thus described by $\tilde{a}_1, \tilde{a}_2, \ldots, \tilde{a}_T$. The objective is to determine the sequence of migration controls $x_1, x_2, \ldots, x_T$ that minimises

$$\sum_{t=1}^{T}(\tilde{Y}_t - \tilde{a}_t)'\tilde{M}_t(\tilde{Y}_t - \tilde{a}_t) \tag{8.7}$$

where $\tilde{M}_t$ is a known symmetric positive semi-definite matrix. It is not necessary that $\tilde{M}_t$ be diagonal. The diagonal elements $\tilde{M}_{ii,t}$ measure the importance of attaining the desired population level $\tilde{a}_{it}$. The desire to account for the interdependencies between regions in meeting the targets will imply the existence of off-diagonal terms in $\tilde{M}_t$. We may then interpret the element $\tilde{M}_{ij,t}$ as the measure of importance of attaining the population level in region i subject to the influences of distance from the target in region j.

For the backward-linkage model similar interpretations are possible for the constraints (8.5), (8.6) and the objective function (8.7). Thus, for both the forward- and backward-linkage models, the national settlement system planning problem can be given the following more general formulation:

For the linear model

$$Y_t = A_t Y_{t-1} + B_t x_t + b_t, \qquad t = 1, 2, \ldots, T \tag{8.8}$$

$$Y_0 = \bar{Y}_0$$

where Y_t is a $n \times 1$ state vector, $\bar{Y}_0$ is a given initial state, x_t is a $m \times 1$ control vector, A_t and B_t are known parameter matrices, the optimal control problem is to determine the set of control variables $x_1, \ldots, x_T$ that minimises

$$W = \sum_{t=1}^{T} (Y_t - a_t)'M_t(Y_t - a_t) \tag{8.9}$$

when the control variables are subject to the constraints

$$F_t x_t = f_t \tag{8.10a}$$

$$R_t x_t \geqslant r_t \tag{8.10b}$$

$$x_t \geqslant 0 \tag{8.10c}$$

for all $t = 1, 2, \ldots, T$ and

$$S_1 x_1 + S_2 x_2 + \ldots + S_T x_T \geqslant s \tag{8.11}$$

In (8.9), a_t is a $n \times 1$ target vector and M_t is a known positive semi-definite matrix. In (8.10b), $R_t = -\tilde{R}_t$, $r_t = -\tilde{r}_t$ and in (8.11) $S_i = -\tilde{S}_i$, $s = -\tilde{s}$.

8.4 *Control theory solutions*

Considerable interest has been directed at finding the optimal control settings for the national settlement system planning problem of the previous section; i.e. given (8.8), minimise (8.9) subject to (8.10) and (8.11).

For the forward-linkage model, Evtushenko and MacKinnon (1976) present an approach based on the penalty function methods of non-linear programming. However, this yields only approximately optimal solutions. Mehra (1975) describes a branch-and-bound integer programming algorithm to solve the forward- and backward-linkage models. In these attempts no budget constraint is imposed for each time (8.10b) into the problem.

The optimal control solution to the problem (even with (8.10b) present) can, however, be developed immediately from the strategy discussed in Sections 3.2 and 6.3. The strategy is in two parts. First, the problem is solved with only the constraint (8.10a) imposed. Secondly, if the resulting solution is infeasible, then the solution is adjusted for feasibility. The adjustment is carried out as a linear programming problem similar to (6.16) using the dual simplex method.

When uncertainties are present either in the form of unknown parameters or an additive error term in the state equation (8.8), the solution methods of Mehra (1975) and Evtushenko and MacKinnon (1976) cannot be modified to handle such uncertainties. However, the two-part solution strategy of Chapters 3 and 6 lends itself to immediate modification in the presence of such uncertainties (see Tan 1980).

8.5 *Manpower planning as a control problem*

A situation related to controlling national settlement system development where job vacancies are used as a control variable is that of controlling manpower in aggregate. Indeed, the two problems can be related to each other as a hierarchical solution similar to that developed in Chapter 7. The discussion below relates to a problem of manpower planning which is typical to many developing countries: to match the supply and training of new entrants into the labour market to the demand for workers at each time period, and particularly to minimise the deviations of actual labour supply from anticipated demands in each skill or training category. This is a problem which has been previously tackled using forecasting and control solutions by Cameron and Nash (1974) and Holt *et al.* (1960), but the solutions developed below make specific use of the constrained control results developed in earlier chapters.

Consider a country's labour force, including new entrants to the labour pool, to be disaggragated into three main 'education–skill' groups:

grade 1 highly educated (e.g. university/college education);
grade 2 medium educated (e.g. upper secondary schooling);
grade 3 lowly educated (e.g. below lower secondary schooling).

Let y_{ti} be the number of workers in the ith grade at year t, and x_{ti} be new entrants to the ith grade. Two assumptions are made regarding the 'mobility' of the labour force:

(a) There is 'upward mobility' for workers in the system, i.e. workers in the system who are in the lower grade at year t are able to move up to occupy jobs in the higher grade at year $t + 1$.
(b) There is 'downward mobility' for the new labour force entrants into the system; i.e. in some situations, because of the incidence of slack labour demand, labour force entrants from the higher grades are 'forced' to accept jobs in the lower grades.

Next define two matrices A and B whose entries represent the magnitude of these two kinds of mobility. Specifically, the $(i - j)$th element of matrix A, i.e. a_{ij}, denotes the proportion of workers in the jth grade who after one time period are 'promoted' to jobs in the ith grade. Also, the $(i - j)$th element of matrix B, i.e. b_{ij}, denotes the proportion of new labour force entrants from the ith grade who are forced to accept jobs in the lower jth grade. It can easily be seen from these assumptions that A must be an upper-triangular matrix while B must be a lower-triangular matrix. The system state y_{ti}, the number of workers in grade i at time t, is determined by $y_{t-1,i}$, as well as being influenced by those workers who were in the lower grades in the previous year and the new entrants from the higher grades. In other words,

$$\left.\begin{aligned} y_{t1} &= a_{11}y_{t-1,1} + a_{12}y_{t-1,2} + a_{13}y_{t-1,3} + b_{11}x_{t1} \\ y_{t2} &= a_{22}y_{t-1,2} + a_{23}y_{t-1,3} + b_{21}x_{t1} + b_{22}x_{t2} \\ y_{t3} &= a_{33}y_{t-1,3} + b_{31}x_{t1} + b_{32}x_{t2} + b_{33}x_{t3} \end{aligned}\right\} \qquad (8.12)$$

or more compactly,

$$y_t = Ay_{t-1} + Bx_t \tag{8.13}$$

where

$$y_t = \begin{bmatrix} y_{t1} \\ y_{t2} \\ y_{t3} \end{bmatrix}, \qquad x_t = \begin{bmatrix} x_{t1} \\ x_{t2} \\ x_{t3} \end{bmatrix}$$

and

$$A = \begin{bmatrix} a_{11} & a_{12} & a_{13} \\ 0 & a_{22} & a_{23} \\ 0 & 0 & a_{33} \end{bmatrix}, \qquad B = \begin{bmatrix} b_{11} & 0 & 0 \\ b_{21} & b_{22} & 0 \\ b_{31} & b_{32} & b_{33} \end{bmatrix}$$

The first-order linear model (8.13) is in the form of an optimal control state-space structure and is similar to that of Cameron and Nash (1974).

With the use of the control approach we are able to define desired levels of manpower requirements for each grade as targets and then to manipulate the control vector of new labour force entrants to achieve these targets. As before, denote these desired targets at year t by the 3×1 target vector a_t with one target for each state variable.

8.6 *Control of manpower in Malaysia*

In the example developed here a ten-stage control problem is developed for annual planning of manpower requirements from 1981 to 1990, with 1980 as the base year providing the initial conditions. A quadratic objective function of the form (2.15) is employed and constraints are introduced which restrict the values which the control variables may take. These constraints arise because f_t, the number of new labour force entrants to the job market, although large, is not unlimited. Hence, for each year the control vector is manipulated to determine the number of new entrants with the respective educational training so as to attempt to meet the targets set. Thus, we have to ensure that allocations do not exceed labour supply, i.e.

$$x_{1t} + x_{2t} + x_{3t} = f_t,$$

or

$$F_t x_t = f_t, \quad t = 1, \ldots, 10 \tag{8.14}$$

where F_t is a 1×3 row vector with all entries equal to unity. This constraint ensures that more workers cannot be put into the labour pool than the supply available.

The solution to this constrained control problem can be obtained in exactly the same way as discussed in earlier chapters and the optimal control settings x_t^* are given at each period by

$$x_t^* = \hat{G}_t y_{t-1} + \hat{g}_t, \qquad t = 1, \ldots, T$$

with feedback matrix $\hat{G}_t$ and forcing vector $\hat{g}_t$ given by

$$\hat{G}_t = G_t - D_t^{-1}F_t'P_t^{-1}F_tG_t$$
$$\hat{g}_t = g_t - D_t^{-1}F_t'P_t^{-1}F_tg_t + D_t^{-1}F_t'P_t^{-1}f_t,$$

where

$$G_t = -D_t^{-1}B'H_tA$$
$$g_t = D_t^{-1}B'h_t$$

and where

$$H_{t-1} = M_t + A'H_t(A + B\hat{G}_t)$$
$$H_T = M_t$$

and

$$h_{t-1} = M_t a_{t-1} + A'(h_t - H_tB\hat{g}_t)$$
$$h_T = M_t a_T$$

form a pair of matrix and vector Riccati equations.

In the empirical example to be discussed here, it is assumed in the specification of the weighting matrix M_t that deviations from targets from the three grades of manpower needs are equally important and are given by the quadratic weights also used in Chapter 7, as Equation (7.7). The example concerns manpower planning in West Malaysia and derives from Tan and Tan (1980). The data are derived from *Source of labour supply – present and future* (Anon. 1978) and various Malaysian plans and manpower survey reports. For the analysis the vectors of both labour supply and demand are disaggregated into three main educational–skill groups based on the categories available from the manpower survey and the Third Malaysia Plan. These three groups are:

(a) the highly educated, comprising those with college diplomas and university degrees;

(b) the medium educated, comprising those with Lower Certificates of Education, School Certificates, its equivalent and Higher School Certificates;

(c) The lowly educated, comprising those with less than Form Three education.

Using these definitions, the examples given in Table 8.1 use a simulation approach in which a variety of alternative definitions of the A and B matrices are assumed and the effect of these different assumptions is then evaluated. Eight different pairs of A and B matrices are assumed. The initial conditions y_0 (i.e. the given number of workers in the three grades for 1980) are as follows:

grade 1 100 749 persons
grade 2 1 199 400 persons
grade 3 2 938 530 persons.

Table 8.1 Assumed structure for the two system state matrices A and B for the eight simulations.

Case	*A*	*B*
1	$\begin{bmatrix} 0.998 & 0.0 & 0.0 \\ 0.0 & 0.998 & 0.0 \\ 0.0 & 0.0 & 0.998 \end{bmatrix}$	$\begin{bmatrix} 0.993 & 0.0 & 0.0 \\ 0.007 & 0.99 & 0.0 \\ 0.0 & 0.01 & 1.0 \end{bmatrix}$
2	$\begin{bmatrix} 0.998 & 0.0 & 0.0 \\ 0.0 & 0.998 & 0.0 \\ 0.0 & 0.0 & 0.998 \end{bmatrix}$	$\begin{bmatrix} 1.0 & 0.0 & 0.0 \\ 0.0 & 1.0 & 0.0 \\ 0.0 & 0.0 & 1.0 \end{bmatrix}$
3	$\begin{bmatrix} 0.996 & 0.0006 & 0.0001 \\ 0.0 & 0.996 & 0.001 \\ 0.0 & 0.0 & 0.996 \end{bmatrix}$	$\begin{bmatrix} 0.993 & 0.0 & 0.0 \\ 0.007 & 0.99 & 0.0 \\ 0.0 & 0.01 & 1.0 \end{bmatrix}$
4	$\begin{bmatrix} 0.996 & 0.0006 & 0.0001 \\ 0.0 & 0.996 & 0.001 \\ 0.0 & 0.0 & 0.996 \end{bmatrix}$	$\begin{bmatrix} 1.0 & 0.0 & 0.0 \\ 0.0 & 1.0 & 0.0 \\ 0.0 & 0.0 & 1.0 \end{bmatrix}$
5	$\begin{bmatrix} 0.998 & 0.0006 & 0.0001 \\ 0.0 & 0.998 & 0.001 \\ 0.0 & 0.0 & 0.998 \end{bmatrix}$	$\begin{bmatrix} 0.993 & 0.0 & 0.0 \\ 0.007 & 0.99 & 0.0 \\ 0.0 & 0.01 & 1.0 \end{bmatrix}$
6	$\begin{bmatrix} 0.998 & 0.0006 & 0.0001 \\ 0.0 & 0.998 & 0.001 \\ 0.0 & 0.0 & 0.998 \end{bmatrix}$	$\begin{bmatrix} 1.0 & 0.0 & 0.0 \\ 0.0 & 1.0 & 0.0 \\ 0.0 & 0.0 & 1.0 \end{bmatrix}$
7	$\begin{bmatrix} 0.992 & 0.0006 & 0.0001 \\ 0.0 & 0.992 & 0.001 \\ 0.0 & 0.0 & 0.992 \end{bmatrix}$	$\begin{bmatrix} 0.993 & 0.0 & 0.0 \\ 0.007 & 0.99 & 0.0 \\ 0.0 & 0.01 & 1.0 \end{bmatrix}$
8	$\begin{bmatrix} 0.992 & 0.0006 & 0.0001 \\ 0.0 & 0.992 & 0.001 \\ 0.0 & 0.0 & 0.992 \end{bmatrix}$	$\begin{bmatrix} 1.0 & 0.0 & 0.0 \\ 0.0 & 1.0 & 0.0 \\ 0.0 & 0.0 & 1.0 \end{bmatrix}$

Table 8.2 Total number of new labour force entrants forecast 1981–1990 which act as a total non-negativity constraint.

Year	f_t
1981	123 515
1982	127 114
1983	130 818
1984	134 631
1985	138 543
1986	129 330
1987	132 748
1988	136 257
1989	139 858
1990	143 566

The number of new labour force entrants f_t from 1981 to 1990 are given in Table 8.2. In the following series of tables (Tables 8.3–8.8) the target vectors and the computed results for the control and state vectors are presented for the eight cases.

The empirical results for the eight cases indicate that the two specifications used for the B matrix do not have significant effect on the calculated state vectors, although the variations are somewhat greater for the optimal control settings (x_t^*). However, the various specifications used for the A matrix do produce more variation in the calculated states and control settings. There is larger variation for y_{t3}, i.e. the number of workers in grade 3. Variations in the calculated y_{t1} are smallest for all eight cases. Furthermore, all the state variables are increasing in magnitude over the years while for the control settings only x_{t1} and x_{t2} show an increasing trend; x_{t3} decreases.

The effects of the attrition rate seem to be interesting. A 0.8% rate of attrition (i.e. the rate of workers leaving the system completely) causes the demand for labour to exceed supply in the last four years of the planning horizon. When the attrition rate is 0.4%, this situation occurs only for the last year, i.e. 1990. The other attrition rates used result in a reverse situation, i.e. labour supply exceeds demand.

It can be seen in Table 8.3 that the number of workers in the highly educated group is increasing from about 102 000 persons to about 204 000 persons between 1981 and 1990, which is consistent with the targeted labour demands. In Table 8.4 the number of workers in the medium educated group is increasing but the deviations from the targeted demands are narrowing over the years, except for the case where the attrition rate is highest. Table 8.5, which refers to workers in the lowly educated group, shows a similar pattern as in Table 8.4.

Before analysing the results of the control settings, attention is drawn to the occurrence of a big jump in the control settings from 1981 to

Table 8.3 Targets and computed results for grade 1 (highly educated) workers.

Year	*Targets*	*(1)*	*(2)*	*(3)*	*(4)*	*(5)*	*(6)*	*(7)*	*(8)*
1981	101 974	102 263	102 263	102 259	102 259	102 269	102 270	102 240	102 240
1982	110 143	110 456	110 456	110 446	110 446	110 468	110 468	110 402	110 402
1983	118 965	119 301	119 301	119 282	119 282	119 319	119 319	119 209	119 209
1984	128 494	128 849	128 850	128 821	128 821	128 876	128 876	128 713	128 713
1985	138 786	139 157	139 157	139 117	139 117	139 194	139 194	138 965	138 965
1986	149 903	150 259	150 259	150 204	150 204	150 307	150 307	150 000	150 000
1987	161 910	162 235	162 234	162 162	162 162	162 297	162 297	161 896	161 896
1988	174 870	175 152	175 152	175 059	175 059	175 231	175 231	174 720	174 720
1989	188 887	189 083	189 083	188 965	188 965	189 183	189 182	188 541	188 540
1990	204 019	204 105	204 105	203 959	203 959	204 228	204 228	203 433	203 433

Table 8.4 Targets and computed results for grade 2 (medium educated) workers.

Year	*Targets*	*(1)*	*(2)*	*(3)*	*(4)*	*(5)*	*(6)*	*(7)*	*(8)*
1981	1 188 923	1 228 222	1 228 222	1 227 610	1 227 608	1 228 928	1 228 927	1 224 974	1 224 974
1982	1 257 547	1 298 348	1 298 348	1 296 923	1 296 924	1 299 740	1 299 743	1 291 304	1 291 304
1983	1 330 133	1 372 070	1 372 072	1 369 712	1 369 709	1 374 264	1 374 259	1 360 664	1 360 669
1984	1 406 908	1 449 512	1 449 510	1 446 071	1 446 074	1 452 599	1 452 608	1 433 121	1 433 123
1985	1 488 115	1 530 778	1 530 779	1 526 106	1 526 111	1 534 904	1 534 901	1 508 729	1 508 727
1986	1 574 009	1 613 201	1 613 202	1 607 135	1 607 129	1 618 484	1 618 489	1 584 743	1 584 742
1987	1 664 860	1 699 132	1 699 135	1 691 474	1 691 480	1 705 723	1 705 727	1 663 469	1 663 477
1988	1 760 956	1 788 598	1 788 598	1 779 168	1 779 169	1 796 656	1 796 654	1 744 887	1 744 891
1989	1 862 598	1 881 598	1 881 597	1 870 179	1 870 178	1 891 285	1 891 283	1 828 922	1 828 921
1990	1 970 027	1 978 045	1 978 045	1 964 407	1 964 412	1 989 537	1 989 541	1 915 422	1 915 427

Table 8.5 Targets and computed results for grade 3 (lowly educated) workers.

Year	*Targets*	*(1)*	*(2)*	*(3)*	*(4)*	*(5)*	*(6)*	*(7)*	*(8)*
1981	2 804 553	3 023 231	3 023 232	3 019 320	3 019 322	3 026 470	3 026 471	3 005 020	3 005 022
1982	2 853 297	3 063 317	3 063 317	3 055 595	3 055 594	3 069 923	3 069 920	3 027 020	3 027 024
1983	2 902 887	3 102 625	3 102 622	3 091 073	3 091 077	3 112 561	3 112 565	3 048 344	3 048 344
1984	2 953 339	3 141 077	3 141 077	3 125 707	3 125 706	3 154 332	3 154 326	3 068 955	3 068 960
1985	3 004 668	3 178 607	3 178 604	3 159 419	3 159 418	3 195 119	3 195 127	3 088 825	3 088 836
1986	3 056 889	3 204 715	3 204 715	3 181 726	3 181 734	3 224 452	3 224 451	3 097 515	3 097 524
1987	3 110 018	3 229 619	3 229 619	3 202 883	3 202 882	3 252 501	3 252 502	3 105 338	3 105 340
1988	3 164 070	3 253 311	3 253 312	3 222 858	3 222 865	3 279 249	3 279 258	3 112 321	3 112 327
1989	3 219 061	3 275 805	3 275 807	3 241 702	3 241 711	3 304 708	3 304 718	3 118 538	3 118 547
1990	3 275 054	3 297 211	3 297 209	3 259 529	3 259 533	3 328 975	3 328 980	3 124 148	3 124 153

Table 8.6 Control settings for grade 1 (highly educated) workers.

Year	*(1)*	*(2)*	*(3)*	*(4)*	*(5)*	*(6)*	*(7)*	*(8)*
1981	1 728	1 716	906	900	713	709	1 292	1 283
1982	8 457	8 397	7 610	7 557	7 415	7 363	8 001	7 945
1983	9 129	9 065	8 253	8 195	8 042	7 985	8 674	8 613
1984	9 857	9 788	8 948	8 885	8 721	8 660	9 402	9 336
1985	10 640	10 565	9 699	9 631	9 455	9 388	10 187	10 116
1986	11 460	11 380	10 485	10 412	10 223	10 151	11 010	10 932
1987	12 363	12 276	11 357	11 277	11 075	10 997	11 919	11 836
1988	13 335	13 242	12 296	12 210	11 994	11 911	12 901	12 811
1989	14 383	14 281	13 311	13 217	12 987	12 895	13 958	13 860
1990	15 509	15 401	14 404	14 303	14 057	13 959	15 097	14 992

1982. To some extent this may be due to inaccuracies in the projections of the data for the initial year of 1980. With this in mind, Table 8.6 shows that, of the new entrants to the labour force for the period 1982–90, a very rapid increase in the highly educated category is required to meet the desired targets (i.e. from about 8000 persons in 1982 to about 14 000 persons in 1990). The required increase for grade 2 workers, although not as rapid as for grade 1 workers, is also evident (i.e. from about 72 000 persons in 1982 to about 99 000 persons in 1990). This is shown in Table 8.7. Finally, Table 8.8 shows a decrease

Table 8.7 Control settings for grade 2 (medium educated) workers.

Year	*(1)*	*(2)*	*(3)*	*(4)*	*(5)*	*(6)*	*(7)*	*(8)*
1981	31 525	31 221	30 367	30 068	29 277	28 988	32 548	32 232
1982	73 257	72 583	71 871	71 208	70 902	70 249	73 809	73 126
1983	77 025	76 321	75 620	74 918	74 745	74 047	77 379	76 670
1984	80 927	80 183	79 479	78 754	78 699	77 986	81 040	80 292
1985	84 940	84 169	83 462	82 696	82 821	82 045	84 782	84 001
1986	86 269	85 486	84 750	83 964	84 225	83 464	85 778	84 998
1987	89 972	89 160	88 392	87 598	88 056	87 252	89 117	88 317
1988	93 709	92 863	92 094	91 254	91 929	91 087	92 456	91 618
1989	97 452	96 576	95 771	94 904	95 812	94 945	95 743	54 878
1990	101 112	100 212	99 361	98 474	99 629	98 737	98 898	98 020

Table 8.8 Control settings for grade 3 (lowly educated) workers.

Year	*(1)*	*(2)*	*(3)*	*(4)*	*(5)*	*(6)*	*(7)*	*(8)*
1981	90 263	90 580	92 243	92 548	93 526	93 819	89 674	90 002
1982	45 402	46 133	47 634	48 349	48 798	49 502	45 503	46 044
1983	44 665	45 433	46 945	47 706	48 031	48 787	44 767	45 536
1984	43 049	44 661	46 204	46 994	47 210	47 987	44 189	45 005
1985	42 964	43 810	45 382	46 216	46 268	47 111	43 574	44 428
1986	31 603	32 468	34 098	34 955	34 882	36 715	32 543	33 400
1987	30 415	31 314	33 002	33 876	33 619	34 501	31 713	32 597
1988	29 214	30 153	31 867	32 796	32 334	33 261	30 902	31 830
1989	28 026	29 003	30 778	31 738	31 060	32 020	30 160	31 121
1990	26 947	27 954	29 801	30 790	29 881	30 872	29 571	30 555

in new labour force entrants from the lowly educated group; from about 46 000 individuals in 1982 to about 28 000 individuals in 1990. This decrease, however, does not bring about a shortage of workers belonging to this group, mainly because the demand for them is not increasing as rapidly as for the other two groups.

It should be clear from the results that the attrition rate is an important factor, determining whether there will be an excess labour supply or a labour demand situation. If the rate of workers leaving the system completely (e.g. through deaths, out-migration or premature and early retirements) exceeds 0.8%, the economy may be faced with the prospect of manpower shortages by the second half of this decade. Even with an attrition rate as low as 0.4%, manpower shortages are likely at the end of the decade, despite any allowance for upward and downward mobility of labour.

Judging from the trend towards rapid reduction in labour force entrants with low levels of education, it is foreseeable that if the planning horizon is extended, shortages will also occur in this category. This leads us to suggest that, in the desire to upgrade the quality of the future labour force by training and education, one should not disregard the economic importance of workers with low education.

9 *Prospects*

9.1 Summary

The past 20 years have seen numerous theoretical developments of optimal control algorithms and their application to practical problems. The field where most effort has been placed, and thus where most progress has been made, has been with engineering problems of industrial plant processes. Perhaps the most dramatic and successful achievement of optimal control methods was the 'Man on the Moon' space programme. Although achievements of such magnitude may not be possible in applications to economic systems, the results available are rather encouraging. Macro-economic applications have shown great potential and it may be said that optimal control methods have become an accepted tool for such economic analysis. More recently, important spatial problems have been given similar optimal control treatment.

In this book a number of new problems have been considered in which optimal control theory can be applied to spatial systems. The early part of the discussion has developed a set of theoretical solution strategies which permit optimal control results in a wide variety of situations; for example, with equality and inequality constraints on the control variables, with constant but unknown parameters, with time-varying parameters, and with constraints in both the state and control variables. This theoretical discussion has been reinforced by simple examples calculated in the text and then amplified in Part II of the book by the solution of a number of practical and large-scale control problems.

It would be foolish to pretend that a solution is offered to all practical or theoretical problems in optimal control of spatial systems. However, the methods discussed in the book should allow a very broad range of problems to be tackled which are tailored to the information available (both *a priori* theory and raw statistical data) in many practical instances. Despite this applicability, a brief discussion is given below of directions in which further research is required in order to extend the range of practical situations which can be treated.

9.2 Research priorities

The problems discussed in Chapters 7 and 8 are already large in scale and have necessitated solution by hierarchical methods. There are, however, a wide range of hierarchical control strategies additional to the ones described in this book. Particularly useful methods may be the aggregation approach (Aoki 1976) or the multilevel approach (Mesarovic *et al.* 1970). In the aggregation approach the system to be controlled is replaced by one which is of lower dimension. The states of the new system are related linearly to the states of the original system by a coupling matrix. The essence of the approach lies in the choice of this matrix. The multilevel approach decomposes the large scale system into several subsystems of lower dimension. It is usually not sufficient to obtain the solution by just solving each of these smaller systems because of the interconnections between them. A second level of control to coordinate the solutions from the first level of subsystems is normally necessary. Furthermore, as we have seen in Chapter 7, it may be necessary to iterate between the first- and second-level problems.

Another area where further research may be fruitful concerns the form of the constraints imposed. In this book all the constraints which have been introduced are deterministic. For example, within the RSG problem, the linear equality constraint ensures that the sum of the grant allocations equals what is made available for distribution. This 'cash limit' for each of the planning stages is assumed to be fixed and known precisely at the start of the planning horizon. It may seem more realistic to relax this requirement by allowing the future cash limits to be random variables. In other words, they are not precisely known, except possibly for the present-time cash limit. In a situation of great uncertainties in government and economic planning, such 'probabilistic' constraints are likely to be more realistic. Methods have been suggested for dealing with such constraints for the static linear programming problem (see, for example, Vajda 1972). These may be classified into two categories – the 'wait-and-see' approach or the 'here-and-now' approach. As the names imply, in the former approach the decision makers wait until the random variables have been observed and then solve the deterministic problem. For a multistage optimisation problem this may not be appropriate. In contrast, the second approach determines the solution now, taking into account the uncertainties associated with the constraints. This appraoch may be useful for dynamic systems, and both methods deserve future attention.

Another possible mathematical extension of the control problem is the introduction of discrete control variables, that is control variables that are restricted to integer values. This might be appropriate in situations where special 'aid' programmes are only made available to some of the spatial areas within the system of interest. The control variable would then have the value of one or zero depending on whether the area is to receive or is not to receive special aid. The presence of integer decision variables in the optimisation of static linear objective function has been extensively studied and is usually referred to as integer linear programming problems. Algorithms have been developed to solve such problems, and an extensive list may be found in Hadley (1964) or Hillier and Lieberman (1974). However, it remains to be seen whether these methods can be modified to tackle integer control variables in the dynamic situation required by optimal control.

These are but a few of the areas where more theoretical advances might be achieved in due course. They will provide the opportunity to enrich the control models developed in the practical application discussed in Chapters 7 and 8, and others that might be conceived. At the same time, empirical research is required to assess the sensitivity of these developments to the problems arising from poor data and other information limitations. It is the authors' fervent hope that the book will stimulate research in each of these areas.

Appendix Notation

This appendix lists notation as used in Part I.

a_t	target vector
W	control performance index
A_t	system state matrix in state-space representation
B_t	system input matrix in state-space representation
Y_t	state vector
x_t	control vector
M_t	weighting matrices in quadratic objective function
T	time horizon
G_t	feedback matrix in unconstrained optimal control rule
g_t	forcing vector in unconstrained optimal control rule
H_t	Riccati matrix
h_t	Riccati vector
x_t^*	optimal control vector
u_t	state-space system noise
v_t	measurement system noise
R	covariance matrix of system noise
z_t	measurement vector
E_t	expectation conditional on all information to end of period t
Φ	Kalman gain matrix
F_t	known coefficient matrix defining equality constraints on controls
f_t	known vector in equality constraints on controls
L	Lagrangian
$\hat{G}_t$	feedback matrix in equality constraints optimal control rule
$\hat{g}_t$	forcing vector in equality constraint optimal control rule
x_t°	first part solution of non-negativity/inequality constrained control problem
$\bar{B}$	matrix for initial dual simplex tableau for non-negativity constrained control problem
x_t^{e}	first part solution to stochastic non-negativity constrained control problem
x_t^{a}	adjusted solution of stochastic non-negativity constrained control problem
I_N	identity matrix of order N
π	stacked vector of unknown system parameters
$\bar{\pi}$	mean vector of π
Q	covariance matrix of π
Σ	covariance matrix of the errors in a seemingly univariate regression model
ϕ_t	time-varying system parameter vector
$\phi_{t/t-1}$	estimates of parameter vector at time t based on information available at time $t-1$
$V(T-k/k)$	conditional quadratic loss function for remaining $T-k$ periods with information up to period k
$x^{\mathrm{e}}(t/k)$	optimal equality constrained control based on time-varying system model at time k
$x^{\mathrm{a}}(t/k)$	adjusted solution based on dual simplex strategy to non-negativity constrained time-varying system at time k
$x^{\mathrm{ic}}(t/k)$	adjusted solution based on method of imposing constraints to non-negativity constrained time-varying system at time k

$\hat{G}_{t/k}$	feedback matrix in equality constrained optimal control rule for time-varying system
$\hat{g}_{t/k}$	forcing vector in equality constrained optimal control rule for time-varying system
R_t	known coefficient matrix defining inequality constraints on controls
r_t	known vector in inequality constraints on controls
S_t	known coefficient matrix defining inequality constraints on states
s_t	known vector in inequality constraints on states

References

Abdelmalek, N. N. 1978. Solutions of minimum time problem and minimum fuel problem for discrete linear admissible control systems. *Int. J. Systems Sci.* **9**, 849–55.

Ahmed, N. U. and K. L. Teo 1981. *Optimal control of distributed parameter systems*. Amsterdam: North Holland.

Anas, A. 1978. Dynamics of urban residential growth. *J. Urban Econ*. **5**.

Ando, A., A. Norman and C. Palash 1978. On the application of optimal control to large scale econometric models. *TIMS Studs Mgmt Services* **9**, 147–70.

Anonymous 1978. *Source of labour supply – present and future.* Research paper no. 13. Malaysia: Department of Statistics.

Aoki, M. 1976. *Optimal control and system theory in dynamic economic analysis*. Amsterdam: North Holland.

Arora, S. S. and M. Brown 1977. Alternative approaches to spatial autocorrelation: an improvement over current practice. *Int. Reg. Sci. Rev*. **2**, 67–78.

Arrow, K. J. 1970. *Public investment, the rate of return, and optimal fiscal policy*. Baltimore, Md: Johns Hopkins University Press.

Åstrom, K. J. 1970. *Introduction to stochastic control theory.* New York: Academic Press.

Åstrom, K. J. and B. Wittenmark 1971. Problems of identification and control. *J. Math. Anal. Applics* **34**, 90–113.

Åstrom, K. J. and B. Wittenmark 1972. On the control of constant but unknown systems. In *Proceedings of the Fifth International Federation of Automatic Control World Congress*, paper 37.5.

Åstrom, K. J. and B. Wittenmark 1973. On self-tuning regulators. *Automatica* **9**, 185–99.

Athans, M. and D. Kendrick 1974. Control theory and economics: a survey forecast and speculations. *IEEE Trans on Automatic Control* **AC-19**, 518–24.

Barber, G. M. and B. A. Ralston 1980. The elementary dynamics of road development. *Geog Anal.* **12**, 258–63.

Batty, M. 1971. An approach to rational design. *Archit. Design* **41**, 436–39.

Batty, M. 1974. A theory of Markovian design machines. *Environ. Plann. B* **1**, 125–46.

Batty, M. 1976. *Urban modelling: algorithms, calibration, predictions*. Cambridge: Cambridge University Press.

Bellman, R. E. 1957. *Dynamic programming*. Princeton, NJ: Princeton University Press.

Bennett, R. J. 1975. The representation and identification of spatiotemporal systems: an example of population diffusion in north-west England. *Trans Inst. Br. Geogs* **66**, 73–94.

Bennett, R. J. 1976. Nonstationary parameter estimation for small sample situations: a comparison of methods. *Int. J. Systems Sci.* **7**, 257–75.

Bennett, R. J. 1977. Consistent estimation of nonstationary parameters for small sample situations – a Monte Carlo study. *Int. Econ. Rev.* **18**, 489–502.

Bennett, R. J. 1979. *Spatial time series: analysis, forecasting and control*. London: Pion.

Bennett, R. J. 1980. *The geography of public finance: welfare under fiscal federalism and local government finance*. Methuen: London.

Bennett, R. J. 1981. A hierarchical control solution to allocation of the British Rate Support Grant. *Geogl Anal*. **13**, 300–14.

Bennett, R. J. 1982a. *Central grants to local governments: the political and economic impacts of the Rate Support Grant*. Cambridge: Cambridge University Press.

Bennett, R. J. 1982b. A hierarchical control solution to allocation of the Rate Support Grant using representative needs as targets. *Trans Inst. Br. Geogs* **7**, 163–86.

Bennett, R. J. and R. J. Chorley 1978. *Environmental systems: philosophy, analysis and control*. London: Methuen.

Bennett, R. J. and K. C. Tan 1978. Stochastic control of regional economies. In *Exploratory and explanatory statistical analysis of spatial data*, C. Bartels and R. H. Ketelapper (eds), 245–69. The Hague: Martinus Nijhoff.

Bennett, R. J. and K. C. Tan 1979. Allocation of the UK Rate Support Grant using the methods of optimal control. *Environ. Plann. A* **11**, 1011–27.

Bennett, R. J. and K. C. Tan 1981. Space–time models of financial resource allocation. In *Dynamic spatial models* D. A. Griffith and R. D. MacKinnon (eds), 367–88. Alphen aan den Rijn: Sijthoff and Noordhoff.

Bernussou, J. and A. Titli 1982. *Interconnected dynamical systems: stability, decomposition and decentralisation*. Amsterdam: North Holland.

Bertsekas, D. P. 1976. *Dynamic programming and stochastic control*. New York: Academic Press.

Blechman, B. M., E. M. Gramlich and R. W. Hartman 1975. *Setting national priorities: the 1976 budget*. Washington, DC: Brookings Institution.

Boyle, L. 1966. *Equalisation and the future of local government finance*. Edinburgh: Oliver and Boyd.

Brainard, W. 1967. Uncertainty and the effectiveness of policy. *Am. Econ. Rev*. **57**, 411–33.

Brito, D. L. 1973. Estimation, prediction and economic control, *Int. Econ. Rev*. **14**, 646–52.

Brock, W. A. 1974. Money and growth: the case of long-run perfect foresight. *Int. Econ. Rev*. **15**, 750–77.

Brock, W. A. 1975. A simple perfect foresight monetary model. *J. Monetary Econ*. **1**, 133–50.

Brown, R. L., J. Durbin and J. M. Evans 1975. Techniques for testing the constancy of regression relationships over time. *J. R. Statist. Soc. B* **37**, 149–92.

Bryson, A. E. and Y.-C. Ho 1969. *Applied optimal control.* Waltham, Mass.: Blaisdell.

Burness, H. S. 1976. On the taxation of nonreplenishable natural resources. *J. Environ. Econ. Mgmt* **3**, 289–311.

Burnett, T. D. and D. Guthrie 1970. Estimation of stationary stochastic regression parameters. *J. Am. Statist. Assoc*. **65**, 1547–53.

Butkovskii, A. G. 1969. *Distributed control systems*. New York: Elsevier.

Caines, P. E. and K. D. Wall 1972. *Theoretical foundations for methods of estimation and control in economic policy optimisation*. Discussion paper No. 3, Programme of Research into Econometric Methods, University of London.

Calvo, G. A. 1978a. On the time consistency of optimal policy in a monetary economy. *Econometrica* **46**, 1411–28.

Calvo, G. A. 1978b. Optimal seignorage from money creation. *J. Monetary Econ*. **4**, 503–17.

Cameron, M. H. and J. E. Nash 1974. On forecasting the manpower requirements of an organisation with homogenous workloads. *J. R. Statist. Soc. A* **137**, 200–18.

Canon, M. D., C. D. Cullum, Jr and E. Polak 1970. *Theory of optimal control and mathematical programming*. New York: McGraw-Hill.

Chadwick, G. F. 1971. *A systems view of planning: towards a theory of the urban and regional planning process*. Oxford: Pergamon.

Chan, Y. T. and J. P. Maillie 1975. Extension of a linear quadratic tracking algorithm to include control constraints. *IEEE Trans on Automatic Control* **AC-20**, 801–3.

Chow, G. C. 1972. Optimal control of linear econometric system with finite time horizon. *Int. Econ. Rev.* **13**, 16–25.

Chow, G. C. 1973. The effect of uncertainties on optimal control policies, *Int. Econ. Rev.* **14**, 632–45.

Chow, G. C. 1975. *Analysis and control of dynamic economic systems*. New York: Wiley.

Chow, G. C. 1976a. The control of nonlinear economic systems with unknown parameters. *Econometrica* **44**, 685–95.

Chow, G. C. 1976b. Control methods for macroeconomic policy analysis. *Am. Econ. Rev.* **66**, 340–5.

Chow, G. C. 1977. An approach to the feedback control of nonlinear econometric systems. In *Frontiers of quantitative economics*, M. D. Intriligator (ed.), Vol. 3A, 263–80. Amsterdam: North Holland.

Chow, G. C. 1981. *Econometric analysis by control methods*. Chichester: Wiley.

Chow, G. C. and S. B. Megdal 1978. The control of large-scale nonlinear econometric systems. *IEEE Trans on Automatic Control* **AC-23**, 344–9.

Clark, C. W., F. H. Clarke and G. R. Munro 1979. The optimal exploitation of renewable resource stocks: problems of irreversible investment. *Econometrica* **47**, 25–47.

Cliff, A. D. and J. K. Ord 1973. *Spatial autocorrelation*. London: Pion.

Cliff, A. D. and J. K. Ord 1980. *Spatial processes*. London: Pion.

Coelho, J. D. and A. G. Wilson 1976. The optimum location and size of shopping centres. *Regional Studies* **10**, 413–21.

Committee of Enquiry into Local Government Finance 1976. *Local government finance*, Cmnd 6453. London: HMSO.

Committee on Policy Optimisation 1978. *Report of Her Majesty's Government's Committee on Policy Optimisation*, Cmnd 7148. London: HMSO.

Copas, J. B. 1966. Monte Carlo results for estimation in a stable Markov time series. *J. R. Statist. Soc. A* **129**, 110–16.

Cripps, F. and W. Godley 1976. *Local government finance and its reform*. Cambridge: Department of Applied Economics, University of Cambridge.

Diamond, P. A. and J. A. Mirrlees 1971a. Optimal taxation and public production: 1. *Am. Econ. Rev.* **61**, 8–27.

Diamond, P. A. and J. A. Mirrlees 1971b. Optimal taxation and public production: II. *Am. Econ. Rev.* **61**, 261–78.

Domanski, R. 1973. A general model of optimal growth in a system of regions. *Papers, Regional Sci. Assoc.* **31**, 73–82.

Dorfman, R. 1969. An economic interpretation of optimal control theory. *Am. Econ. Rev.* **59**, 817–31.

Dreyfus, S. E. 1965. *Dynamic programming and the calculus of variations*. New York: Academic Press.

Duncan, D. B. and S. D. Horn 1972. Linear dynamic recursive estimation from the viewpoint of regression analysis. *J. Am. Statist. Assoc.* **67**, 815–21.

Eastwood, D. B. 1978. An adaptive linear expenditure system for state and local governments. *Appl. Econ.* **10**, 279–87.

Edwards, C. T. 1970. *Public finance in Malaya and Singapore*. Canberra: Australian National University Press.

Evtushenko, Y. and R. D. MacKinnon 1976. Nonlinear programming approaches to national settlement system planning. *Environ. Plann. A* **8**, 637–53.

Eykhoff, P. (ed.) 1973. *Identification of systems and parameter estimation*. Amsterdam: North Holland.

Eykhoff, P. 1974. *System identification*. New York: Wiley.

Fair, R. C. 1974. On the solution of optimal control problems as maximization problems. *Ann. Econ. Social Measurement* **3**, 135–54.

Fan, L. T. and C. S. Wang 1964. *The discrete maximum principle*. New York: Wiley.

Farison, J. B. 1967. Parameter identification for a class of linear discrete systems. *IEEE Trans on Automatic Control* **AC-12**, 109.

Farison, J. B., R. E. Graham and R. C. Shelton, Jr. 1967. Identification and control of linear discrete systems. *IEEE Trans on Automatic Control* **AC-12**, 438–42.

Felderer, B. 1975. Optimal allocation of resources to the educational sector in a system of regions. *Environ. Plann. A* **7**, 59–70.

Foldes, L. 1978a. Optimal saving and risk in continuous time. *Rev. Econ. Studs* **45**, 39–65.

Foldes, L. 1978b. Martingale conditions for optimal saving – discrete time. *J. Math. Econ.* **15**, 81–96.

Foster, C. D., R. A. Jackman and M. Perlman 1980. *Local government finance in a unitary state*. London: George Allen & Unwin.

Friedman, B. M. 1973. *Methods in optimization for economic stabilization policy*. North-Holland: Amsterdam.

Fujita, M. 1976. Toward a dynamic theory of urban land use. *Papers, Regional Sci. Assoc.* **37**, 133–65.

Fujita, M. 1980. *A multi-period model of the urban land market under uncertainty*, 2 vols. Working paper 27. Department of Regional Science, University of Pennsylvania.

Fujita, M. 1981. Spatial dynamics of urban land use. In *Dynamic Spatial Models*, D. A. Griffith and R. D. MacKinnon (eds), 404–39. Alphen aan den Rijn: Sijthooff and Noordhoff.

Garbade, K. D. 1976. On the existence and uniqueness of solutions to multi-period linear/quadratic optimal control problems. *Int. Econ. Rev.* **17**, 719–31.

Gregory, D. J. 1980. The ideology of control: systems theory and geography. *Tijdschr. Econ. Soc. Geog.* **71**, 327–42.

Hadley, G. 1962. *Linear programming*. Reading, Mass.: Addison-Wesley.

Hadley, G. 1964. *Nonlinear and dynamic programming*. Reading, Mass.: Addison-Wesley.

Hall, P. 1974. *Urban and regional planning*. Harmondsworth: Penguin.

Hall, P. 1980. *Great planning disasters*. Harmondsworth: Penguin.

Hepworth, N. P. 1971. *The finance of local government*. London: George Allen & Unwin.

Hillier, F. S. and G. J. Lieberman 1974. *Introduction to operations research*, 2nd ed. San Francisco: Holden-Day.

Ho, Y.-C. and P. B. Brentani 1963. On computing optimal control with inequality constraints. *SIAM J. Control* **1**, 319–48.

Holbrook, R. S. 1974. A practical method for controlling a large nonlinear stochastic system. *Ann. Econ. Social Measurement* **3**, 155–76.

Holbrook, R. S. 1975. Optimal policy choice under a nonlinear constraint: an iterative application of linear techniques. *J. Money, Credit and Banking* **7**, 33–49.

Holt, C. C., F. Modigliani, J. F. Muth and H. A. Simon 1960. *Planning production, workforce and inventories*. Englewood Cliffs, NJ: Prentice-Hall.

Hordijk, L. and P. Nijkamp 1977. Dynamic models of spatial autocorrelation. *Environ. Plann. A* **9**, 505–19.

Howarth, M. J. and P. C. Parks 1971. Studies in the modeling and control of a national economy. *Int. J. Systems Sci.* **1**, 311–22.

Isard, W. and P. Liossatos. 1975. Optimal space-time development: a summary presentation. In *Dynamic allocation of urban space*, A. Karlqvist, L. Lundqvist and F. Snickars (eds), 119–36. Farnborough: Saxon House.

Jacobs, O. L. R. 1974. *Introduction to control theory*. Oxford: Clarendon Press.

Jacobson, D. H. and D. W. Mayne 1970. *Differential dynamic programming*. New York: American Elsevier.

Jenkins, J. and M. Rose 1976. Rate support grants and their application to London. *Greater London Intelligence Q*. no. 35, 9–25.

Johnston, J. 1972. *Econometric methods*, 2nd edn. New York: McGraw-Hill.

Kalman, R. E. 1958. Design of self-optimizing control systems. *Trans Am. Soc. Mech. Engrs* **80**, 468–78.

Kalman, R. E. 1960. A new approach to linear filtering and prediction problems. *Trans Am. Soc. Mech. Engrs. J. Basic Engng D* **82**, 35–45.

Kalman, R. E. and R. S. Bucy 1961. New results in linear filtering and prediction theory. *Trans Am. Soc. Mech. Engrs. J. Basic Engng D* **83**, 95–108.

Kamien, M. I. and N. L. Schwartz 1971. Sufficient conditions in optimal control theory. *J. Econ. Theory* **3**, 207–14.

Kamien, M. I. and N. L. Schwartz 1981. *Dynamic optimisation: the calculus of variations and optimal control in economics and management*. Amsterdam: North Holland.

Karlqvist, A., R. Sharpe, D. F. Batten and J. F. Brotchie 1978. A regional planning model and its applications to South Eastern Australia. *Regional Sci. Urban Econ.* **8**, 57–86.

Kendrick, D. A. 1976. Application of control theory of macroeconomics. *Ann. Econ. Social Measurement* **5**, 171–90.

Kendrick, D. 1977. Applications of control theory to macroeconomics. In *Frontiers of Quantitative Economics*, M. D. Intriligator (ed.), Vol. 3A, 239–62. Amsterdam: North Holland.

Kendrick, D. and J. Majors 1974. Stochastic control with uncertain macroeconomic parameters. *Automatica* **10**, 587–93.

Keyfitz, N. 1968. *The mathematics of population*. New York: Addison-Wesley.

Kmenta, J. and R. F. Gilbert 1970. Estimation of seemingly unrelated regressions with autoregressive disturbances. *J. Am. Statist. Assoc.* **65**, 186–97.

Kolmogorov, A. N. 1941. Local structure of turbulence in a non-compressive fluid with very large Reynold's numbers. *Dokl. Acad. Nauk, SSSR* **30**, 299–303.

Kydland, F. E. and E. C. Prescott 1977. Rules rather than discretion: the inconsistency of optimal plans. *J. Polit. Econ.* **85**, 473–91.

Lai, P. W. 1979. *A transfer function approach to short term stochastic models for soil tension in a sandy soil*. PhD thesis, University of London.

Lee, R. 1978. *Application of optimal control theory to urban and regional planning*. Paper presented at the Meeting of the Association of American Geographers, New Orleans, April.

Lee, R. 1982. Optimal government intervention in regional development. *Geogl Anal.* **14**, 124–34.

Lee, R. C. K. 1964. *Optimal estimation, identification and control*. Cambridge, Mass.: MIT Press.

Le Grand, J. 1976. *The distribution of public expenditure: the case of health care*. Sussex University, Department of Economics, Discussion paper.

Le Grand, J. 1978. The distribution of public expenditure: the case of health care. *Economica* **45**, 125–42.

Leonardi, G. 1978. Optimum facility location by accessibility maximising. *Environ. Plann. A* **11**, 1287–305.

Livesey, D. A. 1973. Can macro-economic planning problems ever be treated as a quadratic regulator problem? In *IFAC/IFORS Conference*, 1–14, IEE Conference Papers No. 101. London: Institution of Electrical Engineers.

Lowry, I. S. 1964. *A model of metropolis*. Santa Monica, Calif.: Rand Corporation.

Luenberger, D. G. 1964. Observing the state of a linear system. *IEEE Trans Military Electronics MIL-8*, 74–80.

McCloughlin, J. B. 1973. *Control and urban planning*. London: Faber.

McDonald, A. G., C. G. Cuddeford and E. J. L. Beale 1974. Mathematical models of the balance of care. *Br. Med. Bull.* **30**, 262–70.

MacKinnon, R. D. 1975. Controlling interregional migration processes of a Markovian type. *Environ. Plann. A* **7**, 781–92.

MacRae, E. C. 1975. An adaptive learning rule for multiperiod decision problems. *Econometrica* **43**, 893–906.

Martin, D. H. and D. H. Jacobson 1979. Optimal control laws for a class of constrained linear-quadratic problems. *Automatica* **15**, 431–40.

Marty, A. L. 1968. The optimal rate of growth of money. *J. Polit. Econ.* **76**, 860–73.

Mayhew, L. D. and G. Leonardi 1982. *Resource allocation in multi-level health care systems: benefit maximisation*. Working paper 16, Instituto Ricerche Economico-Sociali del Piemonte, Turin.

Mehra, R. K. 1975. *An optimal control approach to national settlement system planning*. RM-75-58, International Institute for Applied Systems Analysis, Laxenburg, Austria.

Mesarovic, M. D., D. Macko and Y. Takahara 1970. *Theory of hierarchical multilevel systems*. New York: Academic Press.

Musgrave, R. A. 1961. Approaches to a fiscal theory of political federalism. In *Public finances: needs, resources and utilisation*. Princeton, NJ: Princeton University Press.

Musgrave, R. A. and P. B. Musgrave 1976. *Public finance in theory and practice*. New York: McGraw-Hill.

Nicoletti, B., F. Pezzella and G. Raiconi 1975. Optimum allocation of investments in a two-region economy. In *Proceedings of the IFIP Conference on Optimization Techniques*, 834–46. Nice, France.

Nijkamp, P. 1975. Compromise choices in spatial interaction and regional planning models. In *Dynamic allocation of urban space*, A. Karlqvist, L. Lundqvist and F. Snickars (eds), 289–306. Farnborough: Saxon House.

Nijkamp, P. 1976. Economic dynamics, spatial interaction and environmental control. In *Environmental economics*, P. Nijkamp (ed.), Vol. 2, *Methods*, 175–209. Leiden: Martinus Nijhoff.

Nijkamp, P. 1978. Competition among regions and environmental quality. In *Competition among small regions*, W. Buhr and P. Friedrich (eds). Baden-Baden: Nomos.

Nijkamp, P. 1979a. *Conflicts in international ocean management: with special reference to fishery problems*. Discussion paper 1979–10, Department of Economics, Free University, Amsterdam.

Nijkamp, P. 1979b. *Multidimensional spatial data and decision analysis*. New York: Wiley.

Nijkamp, P. 1980. *Environmental policy analysis*. Chichester: Wiley.

Nijkamp, P. 1981. New developments in multidimensional geographical data and policy analysis, In *European Progress in Spatial Analysis*, R. J. Bennett (ed.) 226–42. London: Pion.

Nijkamp, P. and J. Spronk 1980. *Multicriteria analysis: practical methods*. Farnborough: Gower.

Nijkamp, P. and C. Verhage 1976. Cost benefit analysis and optimal control theory for environmental decisions: a case study of the Dollard estuary. In *Environment, regional science and interregional modelling*, M. Chatterji and P. van Rompuy (eds). Berlin: Springer.

Nissen, D. H. 1968. A note on the variance of a matrix. *Econometrica* **36**, 603–4.

Oates, W. E. 1972. *Fiscal federalism*. New York: Harcourt Brace Jovanovich.

Oates, W. E. (ed.) 1976. *The political economy of fiscal federalism*. Farnborough: Saxon House.

Ohno, K. 1978. A new approach to differential dynamic programming for discrete time systems. *IEEE Trans Automatic Control* **AC-23**, 37–47.

Owens, D. H. and R. P. Jones 1978. Iterative solution of constrained differential/algebraic systems. *Int. J. Control* **6**, 957–74.

Paelinck, J. H. P. and Klaassen, L. H. 1979. *Spatial econometrics*. Farnborough: Saxon House.

Pagan, A. 1975. Optimal control of econometric models with autocorrelated disturbance terms. *Int. Econ. Rev*. **16**, 258–63.

Parks, R. W. 1967. Efficient estimation of a system of regression equations when disturbances are both serially and contemporaneously correlated. *J. Am. Statist. Assoc*. **62**, 500–9.

Peterka, V. and K. J. Åstrom 1973. Control of multivariable systems with unknown but constant parameters. In *Identification of systems* P. Eykhoff (ed.), Vol. 2, 535–44. Amsterdam: North Holland.

Phillips, A. W. 1954. Stabilisation policy in a closed economy. *Econ. J*. **64**, 290–323.

Phillips, G. D. A. 1977. Recursions for the two-stage least-squares estimators. *J. Econometrics* **6**, 65–77.

Phiri, P. A. 1979. *Equilibrium points and control problems in dynamic urban modelling*. PhD thesis, University of Leeds.

Phiri, P. A. 1980. Calculation of the equilibrium configuration of shopping facility size. *Environ. Plann. A* **12**, 981–1000.

Pigou, A. E. 1947. *A study in public finance*, 3rd edn. London: Macmillan.

Pindyck, R. S. 1973. *Optimal planning for economic stabilization*. Amsterdam: North Holland.

Pitchford, J. D. and S. J. Turnovsky (eds) 1977. *Applications of control theory to economic analysis*. Amsterdam: North Holland.

Plackett, R. L. 1950. On some theories in least squares. *Biometrica* **37**, 149–57.

Pontryagin, L. S., V. G. Boltyanski, R. V. Gamkrelidge and E. F. Mishchenko 1962. *The mathematical theory of optimal processes*. New York: Wiley–Interscience.

Prescott, E. C. 1972. The multiperiod control problem under uncertainty. *Econometrica* **40**, 1043–58.

Propoi, A. and F. Willekens 1978. A dynamic linear-programming approach to the planning of national settlement systems. *Environ. Plann. A* **10**, 561–76.

Ralston, B. A. 1979. Marketing as diffusion in a single population. *Geogl Anal*. **11**, 165–74.

Raman, M. A. 1963. Regional allocation of investment: an aggregative study in the theory of development planning. *Q. J. Econ*. **77**, 26–38.

Rao, C. R. 1973. *Linear statistical inference and its applications*, 2nd edn. New York: Wiley.

Rausser, G. C. and J. W. Freebairn 1978. Approximate adaptive control solutions to US beef trade policy. *Ann. Econ. Social Measurement* **3**, 177–203.

Rees, P. H. and A. G. Wilson 1977. *Spatial population analysis*. London: Edward Arnold.

Ripper, M. and P. Varaiya 1974. An optimizing model of urban development. *Environ. Plann. A* **6**, 149–68.

Ritch, P. S. 1973. Discrete optimal control with multiple constraints. I. Constraints separation and transformation technique. *Automatica* **9**, 415–29.

Rogers, A. 1968. *Matrix analysis of interregional population growth and distribution*. Berkeley: University of California Press.

Rosen, J. B. 1967. Optimal control and convex programming. In *Nonlinear programming*, J. Abadie (ed.) 289–304. Amsterdam, North Holland.

Rosenberg, B. 1972. The estimation of stationary stochastic regression parameters re-examined. *J. Am. Statist. Assoc.* **67**, 650–4.

Sakashita, M. 1967. Regional allocation of public investment. *Papers, Regional Sci. Assoc.* **19**, 161–82.

Sakashita, M. and O. Kamoike 1973. National growth and regional income inequality: a consistent model. *Int. Econ. Rev.* **14**, 372–82.

Sakawa, Y. and Y. Ueda 1978. The optimal investment for the control of environmental pollution and economic growth. *Int. J. Systems Sci.* **9**, 193–203.

Sargent, T. J. and N. Wallace 1973. The stability of models of money and growth with perfect foresight. *Econometrica* **41**, 1043–5.

Schoeffler, J. D. 1971. On-line multilevel systems. In *Optimization methods for large scale systems, with applications*, D. A. Wismer (ed.), 291–330. New York: McGraw-Hill.

Scott, A. J. 1971. *Combinatorial programming, spatial analysis and planning*. London: Methuen.

Sengupta, J. K. 1970. Optimal stabilization policy with a quadratic criterion function. *Review of Economic Studies* **37**, 127–45.

Shinners, S. M. 1973. *Modern control system theory and application*. Reading, Mass.: Addison-Wesley.

Shoup, D. C. 1970. The optimal timing of urban land development. *Papers, Regional Sci. Assoc.* **25**, 33–44.

Shupp, F. R. 1976. Uncertainty and optimal stabilization policy. *J. of Public Econ.* **6**, 243–53.

Simon, H. A. 1956. Dynamic programming under uncertainty with a quadratic criterion function. *Econometrica* **24**, 74–81.

Singh, M. G. 1981. *Decentralised control*. Amsterdam: North Holland.

Snickars, F. 1978. Convexity and duality properties of a quadratic intraregional location model. *Regional Sci. Urban Econ.* **8**, 5–19.

Sorenson, H. W. 1976. An overview of filtering and stochastic control in dynamic systems. In *Control and dynamic systems*, C. T. Leondes (ed.), Vol. 12, 1–62. New York: Academic Press.

Strejc, V. 1973. Identification and system parameter estimation. In *Identification and system parameter estimation*, P. Eykhoff (ed.). Amsterdam: North Holland.

Tabata, Y. 1976. Regional economic models and continuous programming. *Reg. Sci. Urban Econ.* **6**, 193–204.

Tan, K. C. 1979a. Optimal control of linear econometric systems with linear equality constraints on the control variables. *Int. Econ. Rev.* **20**, 253–8.

Tan, K. C. 1979b. *Constrained control of spatial systems: an application to allocation of the Rate Support Grant*. PhD thesis, University of London.

Tan, K. C. 1980. Solution strategies for national settlement system planning models. *Geogl Anal.* **12**, 68–79.

Tan, K. C. and S. E. Tan 1980. *Optimal control theory and manpower planning in Malaysia*. Paper presented to a Malaysian Economic Association meeting, 'Agenda for the Nation: Malaysia in the 1980s', Penang.

Theil, H. 1957. A note on certainty equivalence in dynamic planning. *Econometrica* **25**, 346–9.

Theil, H. 1958. *Economic forecasts and policy*. Amsterdam: North Holland.

Theil, H. 1964. *Optimal decision rules for government and industry*. Amsterdam: North Holland.

Tinbergen, J. 1937. *An econometric approach to business cycle problems*. Paris: Hermann.

Tse, E. 1974. Adaptive dual control methods. *Ann. Econ. Social Measurement* **3**, 65–83.

Turnovsky, S. J. 1973. Optimal stabilization policies for deterministic and stochastic linear economic systems. *Rev. Econ. Studs* **40**, 79–96.

Turnovsky, S. J. 1977a. Optimal control of linear systems with stochastic coefficients and additive disturbances. In *Applications of control theory to economic analysis*, J. D. Pitchford and S. J. Turnovsky (eds), 293–336. Amsterdam: North Holland.

Turnovsky, S. J. 1977b. *Macroeconomic analysis and stabilization policy*. Cambridge: Cambridge University Press.

Turnovsky, S. J. 1978. Macroeconomic dynamics and growth in a monetary economy: a synthesis. *J. Money, Credit and Banking* **10**, 1–26.

Turnovsky, S. J. and W. A. Brock 1980. Time consistency and optimal government policies in perfect foresight equilibrium. *J. Publ. Economies B* 183–212.

Tustin, A. 1953. *The mechanism of economic systems*. Cambridge, Mass.: Harvard University Press.

Tzafestas, S. G. (ed.) 1982. *Optimisation and control of dynamic operational research models*. Amsterdam: North Holland.

Vajda, S. 1972. *Probabilistic programming*. New York: Academic Press.

Vishwakarma, K. P. 1974. *Macro-economic regulation: a control system incorporating the Kalman filter for the Netherlands*. Rotterdam: Rotterdam University Press.

West, P. A. 1973. Allocation and equity in the public sector: the hospital revenue allocation formula. *Appl. Econ.* **5**, 153–66.

West, P. A. 1979. *Regional equity in the National Health Service in England: an economic analysis of the hospital sector*. PhD thesis, University of York.

Whitehead, P. G. and P. Young 1979. Water quality in river systems: Monte Carlo analysis. *Water Resour. Res.* **15**, 451–9.

Whitehead, P. G., P. Young and P. A. Mitchell 1978. Some hydrological and water quality models in the ACT Region. Paper presented at the Hydrology Symposium, Canberra, September.

Wiener, N. 1949. *Extrapolation, interpolation and smoothing of stationary time series*. New York: Wiley.

Wieslander, J. and B. Wittenmark, 1971. An approach to adaptive control using real time identification. *Automatica* **7**, 211–17.

Williams, H. C. W. L. and A. G. Wilson 1978. Dynamic models for urban and regional analysis. In *Timing space and spacing*: vol. 3, *Time and regional dynamics*, T. Carlstein, P. N. Parker and N. Thrift (eds), 81–95. Edward Arnold: London.

Wilson, A. G. 1970. *Entropy in urban and regional modelling*. London: Pion.

Wilson, A. G. 1974. *Urban and regional models in geography and planning*. New York: Wiley.

Wilson, A. G. 1981. *Geography and the environment: systems analytical methods.* New York: Wiley.

Wilson, A. G. and R. J. Bennett 1984. *Mathematical and statistical theory in human geography and planning*. New York: Wiley.

Wilson, A. G. and M. J. Kirkby 1975. *Mathematics for geographers and planners.* Oxford: Clarendon Press.

Wolfe, P. 1959. The simplex method for quadratic programming. *Econometrica* **27**, 382–98.

Young, P. C. 1978. General theory of modelling for badly defined systems. In *Modelling, identification and control in environmental systems*, G. C. Vansteenkiste (ed.), 103–36. Amsterdam: North Holland.

Zangwill, W. I. 1969. *Nonlinear programming: a unified approach*. Englewood Cliffs, NJ: Prentice-Hall.

Zellner, A. 1962. An efficient method of estimating seemingly unrelated regressions and tests for aggregation bias. *J. Am. Statist. Assoc.* **57**, 348–68.

Zellner, A. and D. S. Huang 1962. Further properties of efficient estimators for seemingly unrelated regression equations. *Int. Econ. Rev.* **3**, 300–13.

Index